Climate Change
As Popular Science

To:

Dr Khalilur Rabbani Bhuiyan

With compliments of the author,
Harun Rashid,

H Rasid

17 September,
2017

Climate Change
As Popular Science

HARUN RASHID

Emeritus Professor, Department of Geography and Earth Science
University of Wisconsin-La Crosse, Wisconsin, La Crosse, WI, USA

Alan Bauld, Freelance Science Writer, Ottawa, Canada

To order additional copies of this book, contact:
Xlibris
1-888-795-4274
www.Xlibris.com
Orders@Xlibris.com
723440

CONTENTS

Dedication ix
Preface xi
Author Biography xv

1 **Introduction** 1

What is climate change as popular science? 1
Birth of climate change controversy...........................2
Uncertainties in climate change...............................4
Objectives of this book......................................6
Theoretical framework for chapter arrangement.................7
References 9

2 **Greenhouse Gases**..**11**

Solar radiation..11
Earth radiation..13
Composition of the atmosphere................................14
Greenhouse effect..15
Radiation budget...19
References 22

3 **Who Are Polluting Our Atmosphere?** **23**

Atmosphere as an open system 23
Pollution by greenhouse gases................................24
Geography of carbon dioxide emissions........................27
Conclusion...32
References 34

4   **What is Global Warming?** **40**

Definitions and related issues 40
Methods of estimating average global temperatures..................41
Historical trends of global warming 43
Samples of geographical variations in global warming 45
Canada.. 46
Bangladesh..49
Saudi Arabia..52
Limitations of this study and concluding comments..................55
References 56

5   **What Is Climate Change?**................................. **62**

Three distinct components of climate change..........................62
Nature of this topic 63
Has the climate of Toronto become warmer in recent years?..... 64
Have summers become warmer in Toronto? 66
Have winters become milder in Toronto? 68
Has Vancouver become rainier in recent years?..........................69
Conclusion..70
References 71

6   **Climate Change Impacts on Bay of Bengal Cyclones**.......... **78**

Basic problems...78
Popular assumptions about climate change impacts on cyclones....79
Have cyclones become more frequent and
more violent in the Bay of Bengal? 80
Nature of Sea Surface Temperature (SST) anomalies..................83
Effects of ENSO and IOD on Bay of Bengal cyclones...............87
Conclusion..88
References 91

7   **Heat Waves in Delhi and Toronto** **97**

Nature of this study...97

Climatic similarities and differences
between Delhi and Toronto...99
What is a heat wave? 101
2015 May heat waves in India...102
Discourse analysis of newspaper
reports on heat waves in Delhi... 103
Heat waves in Toronto... 112
Discourse analysis of newspaper reports
on heat waves in Toronto... 113
Concluding comments 124
References 127

8   **Discourse Analysis of Newspaper Reports
      on Climate Change Politics in USA....................................144**

Climate change controversies...144
Global warming in USA.. 145
Methods of Discourse Analysis.. 149
Results and interpretations of newspaper discourse................. 152
Global warming hypothesis.. 152
Climate change impacts 157
Mitigation 161
Adaptation to climate change 168
Climate change politics and policies...................................... 172
Letters to the editor... 187
Concluding Comments 195
References 197

9   **Climate Change Policies in Canada 207**

Geographical settings of Canada 207
Canada's positions on the Kyoto Protocol 208
Post-Keystone alternate pipelines 208
Canada's coal reserves...209
Federal emission regulation programs....................................209
British Columbia's emission reduction programs 209
Canada's climate change policies 210
References 212

10 **Conclusion**..**214**

    Climate change in public discourse 214
    Climate change science versus social amplification of risk........ 215
    What have we learned from this study? 217
    Concluding planning implications 218
    References 219

Index..221

DEDICATION

"The digital revolution is far more significant than the invention of writing or even of printing." Douglas Engelbart

This book is dedicated to lifelong learners who are now parts of the digital revolution. Most of the contents of this book are based on digital resources from both websites and libraries.

PREFACE

I (HR) am not a climate scientist with formal training (such as a Ph.D.) in climatology. My current research interest in media discourse on climate change is the outcome of many years of teaching undergraduate courses in climatology and scholarly research in climatic hazards, using both survey data and newspaper discourse analysis. My research journeys have taken a meandering path starting from environmental fluvial geomorphology (Ph.D., University of Saskatchewan, Canada, 1975) and a limited number of peer-reviewed journal articles on morphologic impacts of river dams and flood-control channels (channelization projects) to a fairly large number of publications on flood hazards and floodplain management. It is only at the late stage of my academic career that I have developed research interests in two climate change-related topics: (a) climate change impacts on climatic hazards and (b) media discourse on climate change.

The late stage of my career began in 2004 when I joined the Department of Geography and Earth Science at the University of Wisconsin-La Crosse as a faculty member and department head. Prior to that, I taught at Lakehead University, Thunder Bay, Ontario, Canada for 29 years (1975-2004). In 1975, the Department of Geography at Lakehead University had only five faculty members. Because of the shortage of faculty members and despite my formal training in geomorphology, I was asked to teach a course in climatology, in addition to other physical geography courses. In retrospect, it turned out to be a productive assignment because after teaching climatology for many years the basic concepts of climatology became a second nature to me. In 2004, when I moved to the University of Wisconsin-La Crosse, I was again asked to develop and teach a new course on global warming and climate change. Not only did I develop and teach a regular (classroom) course, I was also asked to develop and teach an online course on climate change.

My interest in research using media discourse goes back a long way to my school days when my late father introduced me to newspapers suggesting how I could gain new knowledge from the media. My lifelong dream of publishing a scholarly journal article using media discourse was fulfilled in 2008 when I was able to co-author an article on the Mississippi floodplain with two of my undergraduate students at the University of Wisconsin-La Crosse (Rasid and others 2008). In this study, we combined conventional survey data with discourse analysis of reports in *The La Crosse Tribune* to explore some of the leading floodplain-management issues in the Mississippi floodplain in La Crosse, Wisconsin (Rasid and others 2008). Following my retirement from the University of Wisconsin-La Crosse in 2009, I have continued my scholarly publications on both climatic hazards and media discourse on climatic hazards. So far, I have published several peer reviewed articles and two co-authored books on climatic hazards (Paul and Rashid 2016; Rashid and Paul 2014). Three of the peer-reviewed articles are based on newspaper discourse analysis: (a) flood disasters in the Red River valley, Manitoba, Canada (Rashid 2011a), (b) cyclone disasters on the Bay of Bengal in Bangladesh and Myanmar (Rashid 2011b), and (c) media framing of public discourse on climate change and sea level rise (2011c).

I have written this unusually long preamble to provide context for my current research interests in media discourse. However, I owe this project to initial ideas of such a book to my co-author Alan Bauld. I have known Alan for more than 40 years, from my days in Lakehead University when he showed me some of the details of audiovisual technology for improving my teaching. Regarding our blog entitled Climate Change As Popular Science (CCAPS), Alan helped me with the set-up of the Wordpress blog (see chapter 1 for the web address). My son Zaid B. Rasid, who works in the IT sector, also helped me with the set-up of the blog. Coming back specifically to this book project, Alan reviewed and edited all of the chapters and also contributed to writing several chapters. Special gratitude is due to the Murphy Library of the University of Wisconsin-La Crosse for granting me unlimited access to its digital resources (because of my status as an emeritus professor). Without that access we could not have finished this project.

I would like to express my special gratitude to my wife Mohsina, the source of my inspiration, for providing support throughout this book project despite major diversions from our family life. Our daughter Moona, who is a fabulous teacher and scholar, is a source of inspiration for me to continue scholarly work. Finally, thanks are due to other members of my family (Zaid, Steve, Shaun and Raiyah) for their unconditional support and encouragement throughout this project and to all of Alan's family members, particularly Lynda for supporting Alan throughout this book project.

Harun Rashid, Maple Ridge (Greater Vancouver), BC, Canada, November 2016

AUTHOR BIOGRAPHY

Harun Rashid, Ph.D. (University of Saskatchewan, Canada, 1975) is an Emeritus Professor at the Department of Geography and Earth Science, University of Wisconsin-La Crosse, USA. Earlier he taught at the University of Wisconsin-La Crosse (2004-2009), Lakehead University, Thunder Bay, Ontario, Canada (1975-2004), University of Benin, Nigeria (1981-1982), and University of Dhaka, Bangladesh (1965-1969). Dr Rashid's administrative experience include nearly two and half years of appointment as Acting Associate Vice-President Research at Lakehead University (April 2002-August 2004) and one term as the department Chair at the University of Wisconsin-La Crosse (2004-2007). Dr Rashid has published extensively, authoring nearly 70 peer-reviewed journal articles and book chapters, on such diverse topics as applied fluvial geomorphology, water resources management, floodplain management, choice modeling of floodplain residents' preferences (with the late Professor Wolfgang Haider), natural hazards and disasters, and media discourse on flood hazards and climate change. He is the co-author of two recent books on climate change and climatic hazards in Bangladesh: Paul and Rashid (2016: Elsevier) and Rashid and Paul (2014: Lexington Books). Dr Rashid serves as a member of the editorial boards of four international journals, namely *International Journal of Disaster Risk Reduction, Disasters, Environmental Management*, and *The Arab World Geographer.*

Alan Bauld, M.A. in Education (Central Michigan University, 1993) and B.A. in Geography (Lakehead University, Thunder Bay, Ontario, Canada, 1977) taught many years at the post-secondary level. Currently he educates and consults in the private sector. He has an abiding interest in climatology. He has also an interest in lifelong learning (adult education), especially in environmental issues and natural sciences.

References

Paul, B.K. and Rashid, H. 2016. *Climatic Hazards in Coastal Bangladesh: Non-Structural and Structural Solutions.* Boston, MA and others: Elsevier Science.

Rashid, H. and Paul, B. 2014. *Climate Change in Bangladesh: Confronting Impending Disasters.* Lanham, MD: Lexington Books.

Rashid, H. 2011a. "Interpreting Flood Disasters and Flood Hazard Perceptions from Newspaper Discourse: Tale of Two Floods in the Red River valley, Manitoba, Canada." *Applied Geography* 31 (1): 35-4.

Rashid, H. 2011b. "Interpreting Cyclone Disasters in Bangladesh and Myanmar from Web-Based Newspaper Discourse: Media Framing of Cyclone Vulnerability on the Bay of Bengal Coast." *The Arab World Geographer* 14 (1): 1-32.

Rashid, H. 2011c. "Media Framing of Public Discourse on Climate Change and Sea Level Rise: Social Amplification of Global Warming vs. Climate Justice for Global Warming Impacts". In *Climate Change and Growth in Asia*, edited by M. Hossain, and E. Selvanathan, 232-260. Cheltenham, UK: Edward Elgar.

Rasid, H., Duffy, K. and Steuck, J. 2008. "Floodplain Management in La Crosse, Wisconsin: Newspaper Discourse vs. Floodplain Residents' Preferences." *Focus* 51 (1): 7-16.

Chapter 1

INTRODUCTION

What is climate change as popular science?

We define popular science as interpretations of scientific concepts in plain language (i.e. in non-technical language) for the general audience, who may or may not have a background in science. Climate change as popular science (CCAPS) is, thus, a non-technical interpretation of climate change science, intended for the general audience. We have a blog on this topic under the following web address: https://climatechangepopulardiscourse.wordpress.com/ . . . retrieved on 23 October 2016. *Earlier versions of some of the chapters of this book have been posted on this website as CCAPS blog posts.*

Climate change is a highly complex science involving physics of the atmosphere (meteorology), long-term climatic data (normally studied by climatologists and geophysicists), past geological and paleontological records for interpreting past environments, ice core data for interpreting past concentrations of atmospheric carbon dioxide and other greenhouse gases, and many other sub-fields of science. Most of these concepts are beyond the grasp of the general audience. Using the popular science language, we have therefore written this book for communicating the most basic concepts of climate change to the general audience. Thus, popular science is essentially a method of "science communication" (Simpson 2015). Paraphrasing Simpson's words, our goal is not only to publicize the climate change science but also to provide a critique of the topic. Our hope is that this book should help readers in gaining an improved understanding of the basics of climate change and politics associated with it. Further, we hope that an improved understanding of the climate change debate would motivate some people to take personal responsibilities for reducing their individual carbon foot prints (i.e. they would cut down their energy use and change consumption behavior in a way that would reduce greenhouse gas emissions). As a more ambitious outcome of this approach, some of the readers (electorates) might persuade their political leaders to support legislations for cutting down greenhouse gas emissions.

Birth of climate change controversy

Climate change has emerged as a highly controversial topic because the idea that climate change is largely man-made (anthropogenic global warming) is a contested hypothesis. While the vast majority of scientists accept scientific findings (published in peer-reviewed scholarly journals) that provide evidence of anthropogenic global warming and the resulting climate change, a small minority of scientists challenge the basic assumptions of anthropogenic global warming, attributing global warming largely to natural changes. Since carbon dioxide emissions (plus emissions of other global warming gases) have been implicated in climate change, emission reductions constitute the basic mitigation measure for global warming and climate change. This has economic and political implications. Powerful coal, oil and other industrial interests have resisted any restrictions on emissions. The Republican Party, which is generally pro-business and pro-industries, has large numbers of climate science deniers. Business lobbies rely heavily on the science of climate change denial and their backers in the Republican Party work on their behalf by blocking many emission cut and other climate change legislative initiatives. The Democrats, in contrast, largely subscribe to the anthropogenic global warming hypothesis and have been pushing for emission cut or emission regulations. The politics of climate change have emerged as an important topic in media and public discourse. The general audience may often be confused by contested claims and counter-claims on climate change. One of our goals in this book is to explain the main concepts of climate change so that the readers can make sense of the noisy debates.

The term global warming refers to a scientifically verifiable concept that "increased greenhouse gases cause the Earth's temperature to rise globally" (Houghton 2004, 335). Climate change is a more general term that is now used widely to refer to the assumed impacts of global warming, increasingly implying any discernible changes in global, regional and even local climates. As a verifiable fact, recorded data indicate that the global average temperature has increased by about 1.44°F (0.8°C) over the last 150 years (Seinfield 2011). Most of the mainstream climate science research attributes this increase to a corresponding increase in the atmospheric concentration of greenhouse gases originating from fossil fuel burning. According to the projections by the Intergovernmental Panel on Climate Change (IPCC), if the current annual rates of anthropogenic contributions of carbon dioxide continue, the atmospheric concentration of CO_2 level is likely to double from its preindustrial level of 280 parts per million (ppm) to about 560 ppm by the end of this century. If no measures are undertaken to curb greenhouse gas emissions under a business as usual scenario, the maximum concentration is likely to reach 650 ppm by 2100 (Houghton 2004, 69). Enhanced absorption of solar and thermal radiation (also called longwave infrared earth radiation: see chapter 2) by this powerful greenhouse gas, in turn, is expected to increase Earth's average air temperature by about 2.16°F (1.2°C) by the end of this century (Seinfield 2011; Soon and Baliunas 2003). This estimate is based purely on calculations of warming by absorption of radiation by carbon dioxide gas (a process called *radiative* or *blackbody* warming). The actual warming that would result is considerably larger, likely to be approximately doubled to about 3.6°F (2.5°C), owing to amplification by climate feedbacks (Houghton 2004, 90).

Some of the leading climate change skeptics, notably Idso (1998), Singer (1996 and 1998), Pearce (1997), and Spencer (2008), do not dispute the current global warming trend, but they suggest that the role of anthropogenic greenhouses gases has been exaggerated by overestimating the positive feedback warming effect of water vapor while at the same time underestimating the cooling effect of clouds (O'Hare 2000, 363). In a more radical departure from the mainstream science of climate change, recently Chilinger and others (2009) have questioned the very notion of anthropogenic global warming by employing a set of mathematical formulations based on a thermodynamic model (called the *adabatic* model) to interpret the effect of enhanced greenhouse gases in atmospheric heating. Focusing on the thermodynamic relationship between air temperature and atmospheric pressure in a greenhouse gas-rich atmosphere, these theoretical interpretations excluded explicitly critical feedback effects between the atmosphere and oceans to reach a contentious conclusion that "significant releases of the anthropogenic carbon dioxide into the atmosphere do not change average parameters of the Earth's heat engine and the atmospheric greenhouse effect" (Chilinger and others 2009, 1207). Although most of the contrarian interpretations of global warming and climate change represent similar incomplete explanations of the current trend in global warming by a small minority of scientists, the origin of climate change controversies can be traced back to such conflicting interpretations of the basic science of global warming.

Uncertainties in climate change

What factors drive such conflicting interpretations of the science of climate change? Uncertainties in climate change derive from at least three sources: (a) feedback uncertainty, (b) climate *forcing*, i.e. "a change in some driver (factor) for the temperature of the Earth" (Archer 2007, 129), and (c) model uncertainty. The climate system is difficult to predict precisely because it is made up of a complex set of internal and external components. A large number of feedbacks (at least 20 have been singled out), some of which are positive (reinforcing) while others are negative (stabilizing), have been identified as being able to influence global climate in discernible and uncertain ways (O'Hare 2000). In understanding the global warming forecast "the feedbacks are everything" (Archer 2007, 4). Water vapor, the most voluminous greenhouse gas in Earth's atmosphere (see chapter 2), also provides the most important positive feedback as vapor absorbs large quantities of outgoing thermal radiation, thus warming air further. The ice *albedo* (reflection) feedback due to global warming is also positive as melting of ice reduces reflection resulting in the availability of greater amounts of solar radiation for warming the atmosphere. The cloud feedback is more complex, depending upon the cloud structure (for further details see Table 5.1 in Houghton 2004, 93), but overall it has a cooling effect (a negative feedback) as it reflects greater amounts of incoming solar radiation than the amount it absorbs from the outgoing thermal radiation. The oceans provide another complex set of feedbacks because of their coupling effects with the atmosphere, simultaneously storing energy in their vast water bodies and releasing/exchanging energy between the ocean surface and the atmosphere.

Climate forcing has played perhaps a greater role in the debate on climate change. Some of the climate change skeptics have attributed orbital forcing and solar forcing as the principal drivers of climate change. The *orbital forcing* deals with variations in the Earth's orbit around the Sun. It consists of three geometric features of the orbit: (a) the *precession angle* of the equinox, (b) the *obliquity of the angle of the pole of rotation*, and (c) *the eccentricity of the elliptical orbit* (Archer 2007). Warming occurs when the Earth is nearest to the Sun during its orbital evolution. Based on ice core data from Antarctica and Greenland, some climatologists have postulated cyclic changes in ice ages (i.e. alterations of colder phases of ice ages and intervening warmer interglacial periods) in response to changes in one or more of these orbital characteristics (the so-called Milankovitch theory). The problem with its direct application to the current global warming trend lies in the time scales of the orbital forcing: the precession cycle takes 20,000 years to complete the entire orbital circle; the obliquity and eccentricity cycles operate within 41,000 and 100,000—400,000 years, respectively (Archer 2007, 91-92). The current global warming trend, measured in time scales of decades and centuries, is not compatible with the long-term scales of orbital forcing. Thus, Chilinger and others' (2009) attribution of current global warming to changes in Earth's orbital precession angles seems to be an incomplete explanation without further analysis of the timing of coincidence between the precession angles and the current global warming trend.

Changes in air temperature in response to volcanic eruptions— cooler temperatures due to reflection of solar radiation by volcanic dusts—provide another example of external climate forcing (Latif 2011). However, because of the erratic nature of volcanic eruptions their overall effects on global warming or global cooling seem to be unpredictable. More successful attempts have been made to relate *solar forcing*, i.e. the impact of the 11-year cycle of sunspot activity, to temperature cycles. Extensive studies on this topic have shown nearly perfect match between changes in Earth surface temperature during the 20th century and the length of the sunspot cycle (see, for example, Friss-Christensen and Lassen 1991; Hoyt and Schatten 1997; Calder 1997; Lean and Rind 1998), but the cause-and-effect chain of sunspot activity is rather weak. An indirect process of warming has been postulated through solar winds and cosmic rays and their effects on lowering the amounts of clouds, hence warmer air temperatures. More significantly, the actual range of recorded variations in solar radiation due to sunspot activity is quite minute (plus/minus 0.1 percent) to alter planetary temperature in a significant way (O'Hare 2000, 361). Another problem with the solar forcing is that it is not capable of explaining the recent warming trend. Enhanced greenhouse gas concentrations in the atmosphere seem to provide a more plausible explanation: "No explanation seems to fit the bill except the rise in greenhouse gas concentrations" (Archer 2007, 129).

Model uncertainty occurs because of the difficulties of incorporating different types of feedback effects adequately in climate models. Further, different models may yield different responses (response uncertainties) to the same external forcing as a result of differences in, for example, physical and numerical formulations (Deser and others 2012). Extensive research has been conducted on reducing model uncertainties, including statistical methods of estimating uncertainties (Allen and others 2000; Weaver and Zwiers 2000). The IPCC assessment reports have gradually improved their methods of indicating model uncertainties in climate change projections. Another problem with modeling is the challenge of quantifying the anthropogenic signal (i.e. human influence on climate) in the presence of the background climate noise (i.e. natural fluctuations) (Latif 2011). Sophisticated fingerprint methods maximizing the signal-to-noise ratio have been applied to detect the anthropogenic signal in observations (Latif 2011).

Objectives of this book

In view of the preceding uncertainties and controversies of climate change, the central objective of this book is to present two types of information. The first type deals with some of the basics of climate change that are beyond these controversies. The second type addresses media and public discourse which are the main sources of these controversies. Specific objectives of this book are three-fold:

- First, to review some of the basics of climate change, focusing on greenhouse effect, sources of greenhouse gas pollution, and the meaning and examples of global warming and climate change.
- Second, to provide examples of climate change impacts: one on the Bay of Bengal cyclones and the other on heat waves in Delhi and Toronto.
- Third, to explore climate change politics in USA and Canada.

The examples are somewhat arbitrary, but one of our goals is to demonstrate the nature of media and public discourse that is driving the climate change controversies.

Theoretical framework for chapter arrangement

In previous sections on climate change controversies and climate change uncertainties, we have used slightly more technical language than our intended goal because of the nature of the subject matters (such as feedback effects, climate change forcing, climate change uncertainties, etc.). Otherwise, we have tried to use the language of popular science throughout the rest of the study. Compared to a mostly quantitative and technical approach of physical sciences, we have used a "social constructionist perspective" as the broad theoretical framework for presenting most of the pertinent issues of climate change as popular science. Central to the social construction of climate change issues, similar to many past environmental issues such as acid rain, ozone depletion, dioxin poisoning, desertification, is the idea that the progress of these issues "varies in direct response to successful claims-making by a cast of social actors that include scientists, industrialists, politicians, civil servants, journalists and environmental activists" (Hannigan 2006, 63). Thus, contrary to an assumption that the science of climate change should be a search for truth, aiming at verifiable objective facts, the current debate on the science of climate change provides a classic example of "science as a claims-making activity" (Hannigan 2006 and 1995). At least two types of knowledge claims are evident in the current debate on the science of climate change. First, *cognitive claims* are made by scientists who have been conducting original research on climate science aiming to convert their findings, hypotheses and theories into publicly accredited factual knowledge (Hannigan 1995, 77). Second, *interpretive claims* are "designed to establish the broader implications of the research findings for a non-specialist audience" (Hannigan 1995, 77). The activities of the IPCC scientists provide an example of technical interpretive claims-making as the researchers in this panel act as scientific advisors to the United Nations (through its pertinent bodies, for example, UNFCCC, WMO, UNEP).

Following this introduction (chapter 1), the contents of the remaining chapters fall into one or more of the elements of social constructionist interpretations:

Verifiable objective facts, i.e. assembling scientific information on climate change. These deal mostly with reviews and synthesis of verifiable scientific information (chapters 2, 3, 4 and 5).

Contested scientific claims: In some of the chapters we have pointed out how climate change has been implicated, often uncritically, in such climatic hazards as cyclones and heat waves (chapters 6 and 7).

Competing policy options: Two chapters on climate change politics in USA and Canada deal largely with media and public discourse on climate change policies (chapters 8 and 9).

The media have developed a track record of reporting on different angles of climate change (emphasizing uncertainties of scientific theories on climate change, opposing views among scientists, and so on), competing policy options proposed by opponents (government policy makers versus civil society groups), conflicts between the establishment and the activists, and so on. In the process, the media often tend to de-contextualize the climate change news. The public, relying heavily on the media for their source of news, picks up the incomplete and out of context information on some of the climate change issues. As a result, some of the issues surrounding climate change and climate change politics have become clouded with an emphasis on the rhetoric. One of the goals of this study is to make a contribution on climate change as popular science beyond the rhetoric of public discourse.

References

Allen, M.R., Scott, P.A., Mitchell, J.F.B., Schnur, R. and Delworth, T. 2000. "Quantifying the Uncertainty in Forecasts of Anthropogenic Climate Change." *Nature* 407: 617-620.

Archer, D. 2007. *Global Warming: Understanding the Forecast.* Malden, MA: Blackwell Publishing.

Calden, N. 1997. *Manic Sun: Weather Theories Confounded.* London: Pilkington Press.

Chilinger, G.V., Sorokhtin, O.G., Khilyuk, L. and Gorfunkel, M.V. 2009. "Greenhouse Gases and Greenhouse Effect." *Environmental Geology* 58 (6):1207-1213.

Deser, C., Phillips, A., Bourdette, V. and Teng, H. 2012. "Uncertainty in Climate Change: the Role of Internal Variability." *Climate Dynamics* 38 (3): 527-546.

Friss-Christensen, E. and Larsen, K. 1991. "Length of the Solar Cycle: an Indicator of Solar Activity Associated with Climate." *Science* 254: 698-700.

Hannigan, J. A. 1995. *Environmental Sociology: A Social Constructionist Perspective.* London and New York: Routledge.

Hannigan, J. A. 2006. *Environmental Sociology*, 2nd edition (London and New York: Routledge.

Houghton, J. 2004. *Global Warming: The Complete Briefing*, 3rd edition. Cambridge, UK: Cambridge University Press.

Hoyt, D.V. and Schatten, K.H. 1997. *The Role of the Sun in Climate Change.* Oxford: Oxford University Press.

Idso, S.B. 1998. "CO2-Induced Global Warming: a Skeptic's View of Potential Climate Change." *Climate Research* 10: 69-72.

Latif, M. 2011. "Uncertainty in Climate Change Projections." *Journal of Geochemical Exploration* 110 (1): 1-7.

Lean, J. and Rind, D. 1998. "Climate Forcing by Changing Solar Radiation." *Journal of Climate* 11 (2): 3069-3194.

O'Hare, G. 2000. "Reviewing the Uncertainties in Climate Change Science." *Area* 32 (4): 357-368.

Pearce, F. 1997. "Greenhouse Wars." *New Scientist* 155: 38-43.

Seinfield, J.H. 2011. Insights on Global Warming." *AICHE Journal* 57 (2): 3259-3284.

Simpson, S. 2015. Book Review: Perrault, S.T. 2013. *Communicating Popular Science: From Deficit to Democracy.* New York: Pelgrave. *Technical Communication Quarterly* 24: 196-198.

Singer, S.F. 1996. "Climate Change and Consensus." *Science* 271: 581-582.

Singer, S.F. 1999. *Hot Talk, Cold Science: Global Warming's Unfinished Debate.* Oakland, CA: Independent Institute.

Soon, W. and Baliunas, S. 2003. "Global warming." *Progress in Physical Geography* 27 (3): 448-455.

Spencer, R.W. 2008. *Climate Confusion: How Global Warming Hysteria Leads to Bad Science, Pandering Politicians and Misguided Policies that Hurt the Poor.* New York: Encounter Articles.

Weaver, A.J. and Zwiers, F.W. 2000. "Uncertainty in Climate Change." *Nature* 407: 571-572.

Chapter 2

GREENHOUSE GASES

What are greenhouse gases?

The main objective of this chapter is to explain how earth's atmosphere is heated by absorption of solar radiation (solar energy) and earth radiation by different greenhouse gases. Water vapor is the most basic and most voluminous natural greenhouse gas that keeps our atmosphere relatively warm by absorbing both incoming solar radiation and outgoing earth radiation. In contrast, carbon dioxide is both a natural and an anthropogenic (man-made) greenhouse gas, that is, its atmospheric concentration has been increased significantly by human activities, especially fossil fuel burning and associated urban-industrial emissions. Other greenhouse gases are much smaller in quantity than carbon dioxide but some of them have high global warming potential. To explain the nature of absorption of different forms of radiation energy by different greenhouse gases, it is helpful to review the nature of energy (radiation) exchanges between the atmosphere and the earth surface.

Solar radiation

The Sun emits (releases) two types of energy from its surface. Most of it is in the form of solar radiation (often called *electromagnetic radiation*). The rest is in the form of out-flowing of hot gases (called *plasma*) and *solar wind* (high-energy particles) which interacts with the earth's magnetic field in the upper atmosphere. Solar radiation, often called solar energy in popular science, is the main source of energy for the Earth. It originates from a very hot Sun. The temperature of the Sun's core is about 15 million °K (degree Kelvin). Compared to its core, the Sun surface is much cooler with an average temperature of 5,785°K (9,953°F or 5,512°C). According to the latest NASA estimate, the average temperature of the earth surface is 288°K (15°C or 59°F). Expressed in the Kelvin scale, the sun surface is at least 20 times hotter than the earth surface (5,785°K/288°K = 20).

Let us focus on solar radiation, which is the main form of energy emitted by the Sun. Radiation is the only form of energy that can travel through a vacuum (such as the space which has no mass) without the presence of a substance. Solar radiation travels from the sun surface to the Earth at the speed of light, i.e., at 300,000 km/s (km/s is the short form for kilometer per second). The original amount of solar radiation/ energy remains unchanged until it hits the top of the atmosphere. Where is the top of the atmosphere? Recent satellite explorations have found traces of atmospheric gases thousands of kilometers above the earth surface. However, for all practical purposes the top of the atmosphere is around 60 miles or 100 km above the earth surface because 99.999 percent of the atmospheric masses (weight of gases) occur within this relatively thin layer of the atmosphere (0-100 km).

In describing the nature of solar radiation striking the top of the atmosphere (or just passing through the atmosphere), some of its physical characteristics are relevant here. Solar radiation arrives at the top of the atmosphere as a bunch or ranges of radiation wavelengths. We can use an analogy to describe the ranges of wavelengths, comparing them to a range of water waves starting from tiny wavelets (ripples) to progressively larger and larger waves. The radiation wavelengths are, of course, much smaller than water waves—so much smaller that they are measured as millionth of a meter (called micrometer or micron). In climate science and atmospheric physics, the entire range of radiation wavelengths is called *electromagnetic spectrum*. In popular language, often the term *sunlight* is used to refer to solar radiation. Strictly speaking, this is misleading as the bulk of solar radiation consists of three dominant *radiation bands* (ranges of wavelengths):

- Ultraviolet (wavelengths shorter than that of light): 7%
- Light (visible spectrum): 44%
- Near Infrared (not visible in our naked eye): 48%

Together, these three types of radiation account for 99% of all solar radiation. The remaining 1% includes (a) gamma rays and x-rays (shorter than ultraviolet waves) and (b) far infra-red, microwaves and radio/TV waves (longer than near infrared). The relative proportions of three dominant radiation bands indicate clearly that *sunlight* accounts for only 44% of total solar radiation, whereas nearly one-half (48%) is in the form of invisible near infrared radiation (invisible to our naked eye). To study how each of these radiation bands interacts with the atmosphere, the sizes of their wavelengths are relevant.

- Ultraviolet: 0.1 to 0.4 micrometer (1 micrometer is one-millionth of a meter or 1 thousandth of a mm)
- Light: 0.4 to 0.7 micrometer
- Near Infrared: 0.7 to 4 micrometer

Together, these three components, i.e., 99% of solar radiation, are called short-wave solar radiation, or simply solar radiation to contrast it from earth radiation (long-wave earth radiation).

Earth radiation

The earth radiation, also called *thermal radiation*, includes all radiation that originates from the Earth, i.e. from the land, vegetation, all water bodies including oceans, lakes and rivers, as well as from all forms of atmospheric water content, such as vapor and clouds. Since their wavelengths are longer than 4 micrometer (i.e. the wavelength limit of 99% of solar radiation), earth radiation is also called interchangeably as long-wave infrared radiation or *thermal infrared radiation* (or just thermal radiation). Most of the thermal infrared radiation ranges from 4 micrometer to 50 micrometer. It is a useful convention to remember that solar radiation is in the form of short waves, whereas earth radiation is in the form of long waves. There are several physical laws that explain the differences between short-wave solar radiation and long-wave earth radiation. Without using mathematical equations, we can summarize the central meaning of these laws as follows: hotter is an object, compared to another cooler object, shorter are the wavelengths of its radiation. The sun surface is 20 times hotter than the earth surface. Therefore, it radiates short waves (solar radiation). Conversely, the cooler Earth radiates long waves. This distinction between short waves (solar radiation) and long waves (earth radiation) is critical for analyzing the greenhouse effect because most of the global warming is the result of absorption of long-wave earth radiation (thermal infrared radiation) by greenhouse gases of the atmosphere.

Composition of the atmosphere

This brings us back to the original question: What are the greenhouse gases? Compared to the total volumes of atmospheric gases, the amounts of greenhouse gases are exceedingly small. Then, how do these gases contribute to global warming? We will try to answer this central question by analyzing the nature of these greenhouse gases and their interactions with different radiation bands (wavelengths of radiation). Three gases account for 99.96% of the volumes of the atmosphere:

- Nitrogen: 78.08%
- Oxygen: 20.95
- Argon: 0.93

At least eleven other gases account for the remaining 0.04% of the atmosphere. Many of them do not affect the behaviour of the atmosphere and are therefore not relevant here. Only the following are leading greenhouse gases (in order of their volumes):

- Water vapor: nearly 0-1%
- Carbon dioxide: 0.0399% (399 ppm, or parts per million, in 2014)
- Ozone: varies between nearly 0 and 1,000 ppm
- Methane: 1.7 ppm

Greenhouse effect

Definition

For assessing the relative importance of each of the leading greenhouse gases, let us review the nature of the greenhouse effect. Expressed somewhat simplistically, the greenhouse effect is an analogy as follows: *Our atmosphere is like a greenhouse; it lets in incoming short-wave solar radiation relatively easily but it does not let out a large quantity of the outgoing earth radiation (thermal infrared radiation) because greenhouse gases absorb them quite efficiently.* In this imperfect analogy the emphasis is on transmission of solar radiation through the atmosphere versus absorption of thermal infrared radiation by greenhouse gases. It is an imperfect analogy because several other processes influence the greenhouse effect, especially reflection of solar radiation, atmospheric counter-radiation and atmospheric windows (see below for details).

Dual role of ozone

When solar radiation penetrates the atmosphere its interactions with atmospheric gases are governed by the nature of a specific gas and its interactions with certain radiation bands. Most of the shortwave radiation transmits (passes) through nitrogen (the bulk of the atmosphere) without any significant absorption. This is because nitrogen is not a reactive gas. Oxygen is a different story. The bulk of the atmospheric oxygen is in the form of oxygen molecules, composed of two atoms of oxygen. The ultraviolet radiation interacts with oxygen molecules at upper levels of the atmosphere (mostly 20-40 miles or 30-60 km) in a complex chain of reactions. To begin with, ultraviolet radiation breaks up oxygen molecules momentarily releasing single atoms of oxygen (highly reactive). Recombination of these single atoms of oxygen with other oxygen molecules creates ozone, a molecule composed of three atoms of oxygen. The ozone then settles down concentrating in the middle atmosphere between 10 and 22 miles (15 and 35 km), with the maximum concentration at 14-16 miles (23-25 km). The ozone, in turn, absorbs almost the entire amount of ultraviolet radiation (especially at the 0.2-0.3 micrometer band), protecting all forms of life on the earth surface from lethal doses of this harmful radiation. This is the most important role of ozone.

Absorption of ultraviolet radiation by ozone has dual effects on air temperature. As it increases air temperature at higher altitudes it has a negative (cooling) effect on the surface temperature because it blocks certain bands of incoming solar radiation. Conversely, ozone becomes a problematic greenhouse gas in the lower atmosphere, especially near the ground where it is produced in surface smog by reactions with gases from industrial activity, mostly automobile exhausts. Such lower level ozone (called *tropospheric ozone*) absorbs several bands of outgoing thermal infrared radiation (within the range of 4 to 14 micrometer).

Water vapor as the most basic greenhouse gas

Water vapor is a natural greenhouse gas that absorbs both incoming solar radiation and outgoing thermal infrared radiation quite efficiently. Water vapor absorbs at least eight short-wave bands (i.e., solar radiation bands) between 0.7 and 3.2 micrometer (all within the near infrared radiation, but most efficiently at the end of the near infrared between 2.4 and 3.1 micrometer). Similarly, water vapor absorbs several bands of outgoing thermal infrared radiation (6.3-8 micrometer and greater than 13 micrometer). With an increase in the average atmospheric temperature due to global warming the absolute amount of atmospheric vapor content is likely to increase. However, unlike continued accumulation of carbon dioxide in the atmosphere, atmospheric concentration of vapor cannot continue to increase indefinitely as it is controlled by the feedback of the hydrologic cycle. Thus, an increase in evaporation due to global warming is likely to result in an increase in condensation and rainfall which, in turn, would remove parts of the moisture from the atmosphere. Yet, water vapor is considered as the most important natural greenhouse gas (a) because its total volume is much larger than that of carbon dioxide and (b) because most of the natural greenhouse effect is attributable to absorption of solar radiation and earth radiation by water vapor.

Carbon dioxide as a natural and anthropogenic gas

Like water vapor, carbon dioxide is also a naturally occurring greenhouse gas, which absorbs certain bands of both shortwave solar radiation (at least four near infrared bands between 1.46 and 4 micrometer, especially 2.6-4.2 micrometer) and thermal infrared radiation (at least four bands between 4.8 and 13.3 micrometer and another band between 14 and 16 micrometer). However, it differs from water vapor in its ability to accumulate in the atmosphere. In particular, emissions from burning of fossil fuels and other human activities have been increasing atmospheric concentrations of carbon dioxide steadily for more than two centuries. As a result, the current concentration of atmospheric carbon dioxide level (as of September 2016) has reached an all-time high of 401 parts per million (shortened as ppm), compared to the pre-industrial (1770s) level of 270 ppm. Because of the continued increases in urban-industrial emissions of carbon dioxide, the current global concerns are focused almost entirely on its potential as an anthropogenic greenhouse gas leading to global warming and climate change. One such concern is the magnitude of potential temperature increases due to doubling of carbon dioxide concentrations in the atmosphere, from its pre-industrial benchmark of 270 ppm to 550 ppm (whenever that number would be achieved). In climate science, estimated temperature increases due to doubling of carbon dioxide concentrations in the atmosphere have been labelled as atmosphere's *climate sensitivity*. At the current rate of carbon dioxide emissions, the Fourth Assessment Report of the Intergovernmental Panel on Climate Change (IPCC) has estimated that the atmosphere's climate sensitivity is within the range of 2°C to 4.5°C (3.6°F to 8.1°F). Stated in a more plain language, with a doubling of the carbon dioxide level in the atmosphere the Earth's climate is likely to experience temperature increases ranging from 2°C to 4.5°C (3.6°F to 8.1°F).

Methane as a greenhouse gas with high global warming potential

Methane is another powerful natural and anthropogenic greenhouse gas that is enhanced significantly by human activities, such as mining, use of natural gas, petroleum industry, rice paddies, landfills, etc. Although atmospheric concentration of methane is significantly lower than that of carbon dioxide (1.7 ppm for methane versus 401 ppm for carbon dioxide), the level of methane has doubled during the last 200 years. Perhaps, more significantly, one of the problems with methane is that, compared to its relatively minute concentration in the atmosphere its global warming potential (GWP) is relatively high—at least 20 to 23 times higher than that of carbon dioxide for a 100 year time horizon. The term global warming potential compares the amount of heat trapped by a given mass of a gas (say 1 kilogram of methane) to the amount of heat trapped by a similar mass of carbon dioxide (say 1 kilogram of carbon dioxide). Thus, if carbon dioxide is used as a benchmark greenhouse gas, methane has a much higher potential of global warming (kilogram for kilogram), despite its smaller amounts of atmospheric concentrations. Like water vapor and carbon dioxide, methane absorbs both shortwave radiation (near infrared band at 3.3 micrometer) and thermal infrared radiation (between 6.5 and 7.66 micrometer). Despite its high potential for global warming, it should be stressed that its absorption bands overlap with water vapor.

Other minor greenhouse gases

These include chlorofluorocarbons (CFCs) and nitrous oxide (laughing gas). The CFCs, an entirely man-made gas, absorb thermal radiation in a narrow band. Its current atmospheric concentration is quite small but its main problem lies with its ability to destroy ozone. Most of the CFCs have accumulated in the *stratosphere* (the second atmospheric layer above the *troposphere*) following their releases from hair spray propellants and leakages from insulation materials and refrigerating gases (such as Freon). Subsequent interactions of CFCs with the ultraviolet radiation in the stratosphere strip off chlorine atoms from CFC molecules. Free chlorine atoms, in turn, react with ozone destroying its molecular structure. This process of ozone destruction by chlorine has resulted in the development of the *ozone hole* over Antarctica. The ozone hole is a concern for increased penetration of ultraviolet radiation to the earth surface, especially in high latitude zones of the southern hemisphere (such as southern Chile and Argentina), where increased incidences of skin cancer have been reported. The good news about CFCs is that further industrial uses of this gas have been regulated and reduced significantly following the Montreal Protocol in

1987. Nitrous oxide is a minor natural and anthropogenic greenhouse gas. Yet, its global warming potential is a concern because of its steady increase in atmospheric concentration from natural and agricultural ecosystems, biomass burning and the chemical industry.

Radiation budget

Definition

So far, we have focussed mainly on absorption of certain bands of radiation energy by different greenhouse gases of the atmosphere. Absorption constitutes only one of the components of a complex process of energy exchanges between the atmosphere and the earth surface. The concept of radiation budget may be helpful for analyzing these exchanges in certain order. *The radiation budget starts with the incoming solar radiation (analogous to income in an accountant's budget) which is balanced against loss of outgoing earth radiation (thermal radiation).* The calculations are quite involving even when percentages of radiation units are used, instead of energy units (such as watts per square meter). The budget does not balance in a straight-forward manner because of several feedback effects (atmospheric effects, cloud effects, etc). To simplify calculations, let us use the bare minimum data in percentages, keeping the analysis as descriptive as possible.

Reflection

We can now ask the question again: What happens when solar radiation strikes the atmosphere or starts passing through the atmosphere? The very first thing that happens is reflection. The most basic type of reflection is scattering of light by gas molecules and dust particles in the atmosphere. The proof of scattering of sunlight is the blue color of the sky which results from scattering of the blue light (wavelength of 0.48 micrometer) by oxygen and nitrogen molecules which have slightly smaller wavelengths than that of the blue color. Larger dust particles and aerosols scatter red light and other sky colors at the twilight. Clouds reflect large quantities of sunlight permanently to the space. When sunlight strikes the earth surface, again large quantities are reflected by different types of earth surface materials: by sand, snow, ice, plant leaves, and all types of water bodies, including oceans, lakes, rivers. All types of reflections, including scattering of radiation in the sky, imply permanent losses from the budget, that is, they do not contribute to warming of the atmosphere and the earth surface. In climate science, the combined values of all reflections are referred to as Earth's *albedo*. The approximate value of the Earth's albedo is about 31% (more precisely, within the range of 30-33%). It should be stressed here that all reflections refer to incoming short-wave solar radiation. In contrast, thermal radiation from the earth surface is *radiated* into the atmosphere (not reflected).

Absorption

Besides the loss of sunlight by reflection, another portion of solar radiation that does not reach the earth surface is the amount of absorption of short-wave radiation by certain atmospheric gases. Approximately one-fifth (18-22%) of the incoming solar radiation is absorbed by ozone and water vapor. Now, if we add this portion (18-22%) with Earth's albedo (30-33%), we can conclude that approximately one-half of the incoming solar radiation does not contribute to heating of the earth surface. It then follows that the remaining one-half of the solar radiation is the primary source of energy for heating the earth surface. Recall our definition of the greenhouse effect; we can now provide some additional details for that definition: Our atmosphere is relatively transparent to about one-half of the incoming solar radiation. This amount is transmitted to the earth surface through an *atmospheric window*. The term atmospheric window is an analogy for the radiation band(s) where absorption of radiation is weak or non-existent. The radiation band for light (0.4-0.7 micrometer) is the most obvious atmospheric window for transmission of short-wave solar radiation to the earth surface.

Earth as the second radiating body

A major change in the Earth-atmosphere energy exchanges follows when solar radiation strikes the earth surface. Absorption of solar radiation by the earth surface results in heating of the earth surface. The heated Earth becomes the *second radiating body* besides the Sun (the primary radiating body). However, the Earth, being much cooler than the Sun, starts radiating its energy as long-wave infrared or thermal infrared radiation (as indicated earlier). Most of this outgoing thermal infrared radiation is absorbed by different greenhouse gases of the atmosphere.

Atmosphere as the third radiating body

Absorption of both short-wave and long-wave radiation by the atmosphere has a direct impact on global warming because absorption transforms (changes) radiation energy into heat which, in turn, increases air temperature. The atmosphere is heated by two sources of energy: (a) absorption of short-wave radiation (solar radiation) and (b) absorption of earth radiation (thermal radiation). Being heated in this manner, the atmosphere becomes the *third radiating body*. It re-radiates again thermal infrared radiation both towards the space and towards the earth surface. This re-radiation by the atmosphere is often called *counter-radiation*. The earth surface is thus heated by absorption of two sources of energy: (a) direct solar radiation and (b) counter-radiation from the atmosphere (which is quite large). That is why the total amount of energy handled by the earth surface is greater than 100% of the incoming solar radiation, reinforcing heating of the earth surface. A warmer than expected earth surface radiates greater amounts of energy into the atmosphere (compared to a gasless vacuum). The combination of absorption of certain bands of incoming solar radiation, outgoing earth radiation and atmospheric counter-radiation, thus, creates an energy cycle that keeps our atmosphere warm like a blanket retaining most of the energy radiated from the earth surface. Only a relatively small amount of thermal infrared radiation from the earth surface (about 12%) escapes into the space through another *atmospheric window* (the thermal radiation window between 13 and 15 micrometer).

How is the budget balanced?

The incoming energy at the top of the atmosphere is 100 unit of solar radiation. This amount is balanced by the 100 unit of radiation losses to the space from the Earth and the atmosphere by a combination of short-waves (solar radiation) and long-waves (thermal radiation) as follows:

- Albedo (short-waves) lost to space: 31%
- Thermal radiation from the earth surface lost to space through the atmospheric window: 12%
- Thermal radiation from the atmosphere (from water vapor, clouds, carbon dioxide) lost to space: 57%

The radiation budget is thus balanced (income 100 units minus losses 100 units), but a perfectly balanced budget would result in a much colder Earth than its current thermal condition. Thanks to the greenhouse effect that the Earth's atmosphere is able to retain some additional energy within its relatively thin layer of greenhouse gases even after balancing the overall energy budget of the Earth-atmosphere-space system. Further details of the balancing processes and calculations of relevant units are quite involving. As this book is intended for the general audience, we have summarized the basic processes in the language of popular science, without delving any further into the complicated steps of calculations.

References

Archer, D. 2012. *Global Warming: Understanding the Forecast* (2nd Edition). Hoboken, NJ: John Wiley & Sons.

Barry, R.G. and Chorley, R.J. 2010. *Atmosphere, Weather and Climate* (9th Edition). New York and London: Routledge.

Goody, R.M. and Walker, J.C. G. 1972. *Atmosphere*. Englewood Cliffs, NJ: Prentice-Hall.

Houghton, J. 2004. *Global Warming: The Complete Briefing* (3rd Edition). Cambridge, UK, New York: Cambridge University Press.

Molders, N. and Kramm, G. 2014. *Lectures in Meteorology*. Switzerland: Springer International Publishing.

Rashid, H. and Paul, B. 2014. *Climate Change in Bangladesh: Confronting Impending Disasters*. Lanham, MD, Boulder, CO, New York, Toronto, Plymouth, UK: Lexington Books.

Chapter 3

WHO ARE POLLUTING
OUR ATMOSPHERE?

Atmosphere as an open system

Why are we starting this topic with a claim like "our atmosphere"? This claim is based on an assumption (or more like an assertion) that the atmosphere is a global common property (called "a global commons") that is shared by all citizens of the world. This assumption is rooted in one of the basic physical characteristics of the atmosphere that it is an open system which exchanges energy and matter freely throughout the entire global atmosphere. The exchange of energy takes place through different forms of radiation. Solar radiation is the most basic form of energy for the atmosphere. Earth radiation (also called thermal infrared radiation) is another form of energy that is radiated by the earth surface into the atmosphere. Matters in the atmosphere include gas molecules (such as nitrogen, oxygen, carbon dioxide, ozone and others), water vapor, dust particles, aerosols, and different types of particles from volcanic eruptions. Differences in air temperature from one place to another result in differences in air pressure which generates winds. At the global scale, a number of major wind systems, such as the southeasterlies, the northeasterlies, southwesterlies, and northwesterlies, transport both energy and matter across the latitudes. Thus, colder winds from the polar region are transported to tropical areas, whereas warmer tropical winds are carried to higher latitudes. Obviously, these exchanges of energy and matter do not follow any political boundary. Thus, air pollution resulting from carbon dioxide emissions in the United States or China may reach distant places like Siberia or Greenland or Bangladesh, crossing many countries in its path. Because of such free movement of air pollution throughout the global atmosphere it is not an unreasonable question to ask who are polluting our atmosphere. The answer may already be known in general, but this article analyzes a set of global monitoring data on carbon dioxide emissions by different countries of the world. We will demonstrate that only a limited few industrialized countries are responsible for the bulk of these emissions.

Pollution by greenhouse gases

Any unwanted characteristics of the atmosphere may be considered as air pollution. It could be a bad smell or an undesirable composition of the atmosphere, such as excessive dust particles or excessive concentrations of certain gases. Increasing concentrations of anthropogenic (human-induced) greenhouse gases in the atmosphere, such as carbon dioxide, methane and ozone, may be considered as the worst form of undesirable characteristics of the atmosphere as they contribute to global warming and climate change. In this article, the focus is on carbon dioxide, partly because it is the leading anthropogenic greenhouse gas but mainly because of the availability of a set of data on recent trends in carbon dioxide emissions (see below for further details on data source).

Not all greenhouse gases are harmful for our atmosphere. The most basic greenhouse gas is *water vapor* which absorbs parts of both solar radiation and earth radiation, keeping our atmosphere warmer than expected (compared to an atmosphere without any vapor). Water vapor is essentially natural. Any increase in atmospheric vapor content by human activities is controlled by the feedback of the hydrologic cycle which removes parts of the atmospheric moisture by precipitation. That is why, water vapor is not considered as an anthropogenic gas. Water vapor is also the most voluminous greenhouse gas, that is, its atmospheric concentration is much higher than that of *carbon dioxide* which ranks second to vapor in its concentration. Despite its smaller concentrations in the atmosphere than that of vapor, carbon dioxide is the main concern for global warming because of its ability to accumulate in the atmosphere and because it is both a natural and an anthropogenic greenhouse gas. The natural carbon dioxide is a very important component of the global carbon cycle. The bulk of the atmospheric carbon dioxide is released into the atmosphere from ocean surface (from dissolved carbon dioxide). Similarly, oceans also absorb atmospheric carbon dioxide. Land-based organisms, including all animals and humans, exhale carbon dioxide. In turn, natural vegetation utilizes atmospheric carbon dioxide in a chemical reaction with sunlight (called photosynthesis) to create its biomass (leaves, trunks, branches, etc.). The problem with carbon dioxide lies in its ability to accumulate in the atmosphere due to human activities. This is called anthropogenic or human-induced carbon dioxide. The following is a summary of data for natural and anthropogenic sources of carbon dioxide:

Natural sources of atmospheric carbon dioxide:

- Oceans: 43%
- Plant and animal respiration: 29%
- Soil respiration and decomposition: 28-29%
- Volcanic eruption: negligible percentages

Anthropogenic (human) sources of carbon dioxide:

- Burning of fossil fuels (coal, oil [petroleum], natural gas): 87%
- Land use (deforestation): 9%
- Industrial processes (especially cement production): 4%

Among the fossil fuels, coal is the worst polluter, accounting for about 43% of all fossil fuel emissions. Recently, six of the eight G8 countries (leading industrial nations) produced nearly two-thirds of their electricity by burning coal. The transportation sector (automobiles, railways and ships), burning mainly petroleum (gasoline and diesel) is another major polluter. According to the International Energy Agency (2012), economic activities of industrial societies are heavily dependent on burning fossil fuels, as the following data indicate:

Economic activities dependent on burning fossil fuels:

- Electricity and heat generation: 41%
- Transportation: 22%
- Industry: 20%
- Residential: 7%
- Others: 10%

Like carbon dioxide, *methane* is another greenhouse gas which is both natural and anthropogenic. There are at least two major concerns with methane. First, although its concentration is much smaller than that of carbon dioxide it has doubled in the last two centuries due to human activities. Second, the main concern with this increase is that methane has a much higher global warming potential than carbon dioxide. Each molecule of methane has about eight times higher potential of warming the atmosphere than a carbon dioxide molecule. Wetlands are the largest natural sources of methane emissions (about one-quarter of the total) but a variety of anthropogenic sources contribute about two-thirds of the total emissions. Among anthropogenic sources, mining, gas and petroleum industry produce about one-fifth of the total.

Natural sources of atmospheric methane:

- Wetlands: 27%
- Oceans: 3%
- Others: 6%

Anthropogenic sources of methane:

- Coal mining, natural gas, and petroleum industry: 18%
- Enteric fermentation (gases) from livestock: 16%
- Rice paddies: 11%
- Landfills, waste treatment, biomass burning: 20%

Nitrous oxide, also called laughing gas, is a relatively minor greenhouse gas. Yet, it is a concern for global warming because it has a long life time in the atmosphere. It stays in the atmosphere for at least 100 years. It has also increased by about 16% from the pre-industrial period. It is largely a natural greenhouse gas originating from oceans and forest soils. However, human activities, such as fossil fuel combustion and agriculture, have contributed at least one-third of its present concentration.

Ozone is again another natural greenhouse gas (especially in upper atmosphere), but it becomes a problematic anthropogenic gas in the lower atmosphere (called *tropospheric ozone*) due to its chemical reactions with ground-level pollution, such as smog and automobile exhausts. The global amounts of ozone are difficult to estimate because they vary widely from one place to another.

Geography of carbon dioxide emissions

Data source

The central question of this chapter, "who are polluting our atmosphere", relates directly to the geography of carbon dioxide emissions. When it comes to its impacts on climate change, it has been said that *most of the carbon dioxide emissions originate in highly developed industrial countries, but adverse impacts of climate change resulting from these emissions affect developing countries disproportionately, though they have contributed very little to these emissions.* This is a significant assumption which has two distinct components. The first part deals with geographical disparity in carbon dioxide emissions. The second part is concerned with a more difficult assumption regarding impacts of global warming. Adverse impacts of climate change do not necessarily imply only changes in air temperature and precipitation but also include a host of other related changes, such as increases in natural disasters (floods, droughts, cyclones, etc), impacts on forests, agriculture and glaciers, just to name a few. It is not an easy task to prove potential impacts of climate change on the physical environment. Many variables (factors) are involved and the cause-and-effect relationships are not straight-forward. In contrast, we have a set of data on carbon dioxide emissions for the last 25 years which tell us where from these emissions are coming. This data source is called EDGAR: Emission Data for Global Atmospheric Research, a set of data on carbon dioxide emissions (tables of data) by different countries of the world. These data are monitored by the Joint Research Centre (JRC) of the European Union (EU) and are available on the website: https://edgar.jrc.ec.europe.eu . . .

About the units used in Tables 3.1-3.4

To avoid use of excessive conversions of data, most of my interpretations in this article are based on percentages of the amounts (weights) of carbon dioxide emissions. The basic emission data in original EDGAR tables are in the form of ktons (short-formed for kilotonnes, where 1 kilotonne = 1,000 metric tonnes). In Tables 3.1-3.4, original data for total global emissions in ktons have been replaced with Gt (gigatonnes, where 1 Gt = 1 billion metric tonnes), to avoid long numbers with too many digits. For those who are not familiar with the metric units of weights, the conversion table in the endnotes may be helpful.

The Kyoto Protocol

The 1992 Earth Summit in Rio de Janeiro, Brazil was the beginning of a major global initiative to curb greenhouse gas emissions. The summit was organized by the United Nations and was officially called the United Nations Conference on Environment and Development (UNCED). The conference was so large that, to accommodate nearly 25,000 participants, several simultaneous meetings were organized on different topics. Based on the total numbers of participants and the ranges of representatives—heads of states (116), representatives from 172 governments, thousands of environmentalists and NGOs, businessmen and university research scholars—it was perhaps the largest such conference ever held. An important outcome of the conference was an agreement in principle among participating countries to reduce greenhouse gas emissions under a framework called the United Nations Framework Convention on Climate Change (UNFCCC). Initially, the UNFCCC was ratified by 154 countries. As of 2015, all of the 196 UN member-states have ratified the convention. One of the first tasks of the UNFCCC was to set up a baseline (target) for monitoring future greenhouse gas emissions, to stabilize their levels in the global atmosphere. It was agreed that the developed industrialized countries should take a lead in this initiative. Therefore, in 1997, the Kyoto

Protocol (a climate treaty) was formulated to firm up commitments from a group of industrialized nations (called Annex I countries) for reducing their greenhouse gas emissions by specific percentages from their baseline emissions in 1990. The details of the Protocol were worked out in another UNFCCC Conference of Parties (COP) in 2001, which had set up the first commitment period between 2008 and 2012 and the second commitment period starting from 2012. The EDGAR emission data for 1990, 2000, 2010 and 2013 may thus be used as a report card for global carbon dioxide emissions by Annex I countries (and others).

Leading polluters

The summary data in Table 3.1 demonstrate a continuing and a staggering geographical disparity in carbon dioxide emissions throughout the reporting period (1990-2010). First, despite UNFCCC's efforts to curb carbon dioxide emissions, total global emissions had increased from 22 Gt in 1990 to 25 Gt in 2000 and 33 Gt in 2010. Second, only ten countries continued to contribute about two-thirds of the total global emissions. Of them, six to eight countries were Annex I nations. In 1990 and 2000, the United States led this group, contributing alone about one-quarter of the total global emissions. In 2010, a major shift in this ranking occurred as China, a non-Annex I nation, rose to the top rank contributing at least one-quarter of the total global emissions and outdistancing the United States to the second rank (with 16% of global total). China's rise to the top as the infamous polluter of the global atmosphere has been meteoric in terms of its time line. In 1990, it ranked second with 11.21% of global emissions while Russia ranked third with a very similar contribution (11.06%). In 2000, it outdistanced Russia for the second place, whereas in 2010, it outdistanced the United States for the top rank. The global implications for China's ascension to the top-ranking emitter of carbon dioxide are devastating. With no formal commitment to reduce its greenhouse gas emissions as a non-Annex I nation, in recent years China has accelerated its carbon dioxide emissions into the global atmosphere, negating other global initiatives to curb emissions of this harmful greenhouse gas. Third, in a similar development, another group of eight to nine non-Annex I countries ranked among top 20 contributors of carbon dioxide emissions. Among them, India rose to the third rank in 2010 with

5.38% of global emissions. Others among this group include some of the rapidly industrializing nations, such as South Korea, Mexico, Indonesia and Brazil, but the total contributions by each of these countries ranged between 1% and 2% of the global emissions.

We can look into this staggering geographical imbalance in the source areas of carbon dioxide emissions by comparing emission data for the 20 top-ranking countries with that of the remaining countries of the world. The EDGAR data included several non-sovereign territories, such as Puerto Rico, Guam and some of the British and French possessions. The data also included a limited number of micro-states which contributed insignificant amounts of carbon dioxide emissions (each accounting for less than 0.01% of the global total). For the current study data were analyzed for 187 independent countries which contributed 99.7% of the total global emissions in 1990. As indicated in Table 3.1, only 20 countries were responsible for 77%-80% of the total global carbon dioxide emissions, whereas the remaining 167 countries contributed only 20-23% of the total emissions. Data in Table 3.2 indicate that the latter included not only most of the developing countries of Asia, Middle East, Africa and South America but also many European nations, including nearly two-thirds of the Annex I nations. Among the non-Annex I European countries, Greenland and Gibraltar each contributed less than 0.01% of global emissions. On the upper end of the scale, each of the 20 Annex I countries of Europe contributed 0.1%-0.99% of global emissions. These included the following countries (in alphabetic order): Austria, Belgium, Belarus, Bulgaria, Czech Republic, Croatia, Denmark, Estonia, Finland, Greece, Hungary, Ireland, Lithuania, Netherlands, Norway, Portugal, Romania, Slovakia, Sweden and Switzerland. With an annual emission rate of about 0.06% of the global total, Bangladesh ranked #87, i.e., it belonged to the low emission group countries each of which contributed 0.01%-0.09% of global carbon dioxide emissions.

Carbon dioxide emission reduction challenges for Annex I countries

One of the basic commitments of the Kyoto Protocol was that the Annex I countries were expected to maintain their carbon dioxide emission levels in 2000 up to or below that of the 1990 baseline emission levels. To determine which countries passed this test, the Annex I countries were divided into two groups: (a) countries that exceeded the 1990 emission levels and (b) others who were able to reduce their emissions below the 1990 levels. The differences in emission rates between 2000 and 1990 were converted into percentages of increases or decreases, compared to the 1990 emission rates (Table 3.3). The majority of the Annex I countries (21 out of 39) failed to keep their commitment by increasing their emission rates by as little as 1% to as high as 57%. About one-third of these countries belonged to the top 20 high emitters. These included (in alphabetic order): Australia, Canada, France, Italy, Japan, Spain and USA. Among the 19 countries who managed to reduce their emission levels below that of 1990, only Russia, Germany, Ukraine and Poland belonged to the top 20 high emitters. Thus, the results of this simple test were mixed at best. Some of the Annex I countries succeeded in the test by reducing their emissions significantly, but on balance the major polluters, such as Australia, USA, and Canada continued to increase their carbon dioxide emissions.

Per capita emission rates in 2013

The latest data for 2013 seem to indicate a continuation of the new pattern that was established in 2010 with China contributing almost double the amounts of emissions by the United States (Table 3.4). While India, Russia, Japan, Germany, South Korea and Canada have maintained their relative ranks, Brazil and Indonesia have emerged as the 9th and the 10th top-ranking emitters. UK and Mexico have reduced their emissions marginally and thus have moved down to the group of 20 top-ranking emitters. So far, this type of ranking has been based on the total amount of emissions by each country (in ktons from EDGAR tables). Such national data are important for determining relative contributions of each country to the total global emissions of carbon dioxide. However, from the perspectives of global responsibility perhaps per capita emission is a better reflection of individual contributions to pollution of the global atmosphere (Table 3.4). Expressed in this manner, geographical disparity in carbon dioxide emission sources may be interpreted somewhat differently. Thus, whereas only ten countries (out of 187) contributed at least two-thirds of the total global emissions in 2013, this list included several population giants, such as China, India, USA, Indonesia, Brazil, Russia, and Japan.

Expressed in terms of the global population, in 2013 nearly one-half of the world's population (3.34 billion or about 46% of 7.12 billion) contributed two-thirds of the total global emissions of carbon dioxide. The bulk of the geographical disparity in emission sources is thus attributable to about one-half of the global population. On a further analysis of shared responsibilities, a much more meaningful picture appears when the total amounts of carbon dioxide emissions by each country (expressed in metric tonnes) are divided by its population. With such data (as presented in Table 3.4) we can see that individual responsibilities for carbon dioxide emissions shift again to highly industrialized countries, especially to Australia, USA and Canada, each contributing 16-18 tonnes per person. Considering its modest share of total global emissions (1.36%), the highest per capita emission by Saudi Arabia was a reflection of its relatively small population (27 million). In contrast, the lower per capita emission by China (7.65 tonnes per person) is clearly the reflection of its huge population (1.3 billion). Even if we consider the large population of China in determining its share of global emissions of carbon dioxide, its per capita emission in 2013 still exceeded that of ten countries out of top 20 emitters (Table 3.4).

Conclusion

Coming back to the original question, "who is polluting our atmosphere", the answer is not necessarily straight-forward since so many issues are involved. To start with a simplistic answer, when we think of air pollution it evokes images of car exhausts, especially during traffic congestion. Among other images, we may think of dark pollution plumes from industrial chimneys or power plants. We may even think of narrow bands of exhaust high in the sky from commercial jets or even less threatening exhaust from our home heating. If you ask a casual question about potential sources of air pollution, people are likely to choose one or more of the above sources. If you would press further about potential gaseous composition of air pollution, carbon dioxide is likely to be the top answer, especially if you would provide a multiple choice. Similarly, if you ask a slightly more challenging question, "which countries do you think are responsible for global warming and climate change", the likely answer would point out fingers to highly industrialized nations. In short, it is reasonable to assume that most of the people may be at least vaguely familiar with the subject matter of this article. What is then the contribution of this article? The specific answer is that a comprehensive analysis of the latest EDGAR data—a set of primary (original) data collected by the Joint Research Centre of the European Union—has confirmed or re-confirmed the following assumptions.

- First, according to the Joint Research Centre of the European Union, the main reason for focusing EDGAR data on carbon dioxide is that it is the leading anthropogenic greenhouse gas responsible for global warming.
- Second, conforming to popular perceptions, these data show that, indeed, a limited few industrialized countries are responsible for the bulk of the carbon dioxide emissions. More specifically, this analysis shows that only ten countries (out of a total of 187) have been contributing at least two-thirds of the total global emissions of carbon dioxide for the last quarter of a century (period of analysis). If we expand the list to 20 countries, their contributions increase to about 77-80%.
- Third, since 2010 China has taken over the United States as the top-ranking contributor to carbon dioxide emissions, accounting for nearly one-third (29% in 2013) of the total global emissions. This amount was approximately double that of the United States (15% in 2013).
- Fourth, China's global responsibility for carbon dioxide emissions is moderated somewhat when data are expressed as per capita emissions. Such data implicates again mostly the industrialized Western nations, particularly USA, Australia, Canada and Russia, who lead in per capita emissions. The data, however, do not absolve China's responsibility as its per capita emissions in 2013 still exceeded those of at least ten out of 20 top-ranking emitters.

The main reason for continued increases in carbon dioxide emissions is that fossil fuel combustions associated with economic activities, especially urban-industrial activities, contribute to carbon dioxide emissions, in addition to other greenhouse gases. Since we have become so much accustomed to the quality of modern life that is highly dependent on fossil fuel-based economy, it will not be an easy task to curb greenhouse gas emissions. That is why we have seen continued growth in carbon dioxide emissions even following the Kyoto Protocol commitments to reduce such emissions. National economic interests are at stake here. Ignoring the Kyoto Protocol, China has continued its unabated emissions justifying its right to achieve economic prosperity. India has followed the suit and now ranks third in global emissions. The United States has taken an uncertain and euphemistic position, carefully avoiding Kyoto Protocol commitment but adopting limited emission control measures on its own initiatives. Canada has withdrawn its ratification of the Kyoto Protocol. As the successive post-Kyoto Protocol Conferences of Parties (COP) indicate, the politics and economics of carbon dioxide emissions are so complex that the entire global community seems to be struggling for achieving a balance between economic development and the need to control total amounts of carbon dioxide emissions.

End Notes

Metric units of weights (metric tons are spelled as tonnes, compared to English tons)

- 1 kilogram (kg) = 1,000 grams
- 1 metric tonne (short-formed as t) = 1,000 kg
- 1 kilotonnne (short-formed as kton or kt) = 1,000 tonnes
- 1 megatonne (Mt) = 1,000,000 tonnes (1 million tonnes)
- 1 gigatonne (Gt) = 1,000,000,000 tonnes (1 billion tonnes)

References

Original data (tables of data) analyzed in this chapter were retrieved from a website called EDGAR: Emission Data for Global Atmospheric Research. The site is maintained by the Joint Research Centre (JRC) of the European Union and its web address is: https://edgar.jrc.ec.europe.eu

The original EDGAR report was prepared by:

Olivier, J.G.J., Janssens-Maenhout, G., Muntean, M. and Peters, J.H.A.W. 2014. *Trends in Global CO$_2$ Emissions – 2014 Report.* JRC Report 93171/PBL Report 1490. ISBN 978-94-91506-87-1, December 2014.

Textbooks

Houghton, J. 2004. *Global Warming: The Complete Briefing* (3rd Edition). Cambridge, UK, New York: Cambridge University Press.
Rashid, H. and Paul, B. 2014. *Climate Change in Bangladesh: Confronting Impending Disasters.* Lanham, MD, Boulder, CO, New York, Toronto, Plymouth, UK: Lexington Books.

Table 3.1   Geography of Carbon Dioxide Emissions in 1990, 2000, and 2010

1990 Total global emissions: 22 Gt	2000 Total global emissions: 25 Gt	2010 Total global emissions: 33 Gt
*USA (22.62% of global total)	**USA** (23.13% of global total)	China (26.36% of global total)
China (11.21)	China (13.88)	**USA** (16.37)
Russia (11.06)	**Russia** (6.56)	India (5.38)
Japan (5.27)	**Japan** (5.03)	**Russia** (5.18)
Germany (4.63)	India (4.18)	**Japan** (3.75)
Ukraine (3.49)	**Germany** (3.42)	**Germany** (2.49)
India (2.99)	**Canada** (2.17)	South Korea (1.79)
UK (2.67)	**UK** (2.15)	**Canada** (1.68)
Canada (2.03)	**Italy** (1.81)	**UK** (1.54)
Italy (1.93)	South Korea (1.77)	Mexico (1.39)
Top 10 countries: 68%	Top ten countries: 64%	Top 10 countries: 66%
France (1.78)	**France** (1.61)	Indonesia (1.37)
Poland (1.41)	Mexico (1.49)	Brazil (1.32)
Mexico (1.41)	**Australia** (1.41)	**Australia** (1.31)
Australia (1.24)	**Ukraine** (1.39)	Saudi Arabia (1.29)
South Africa (1.22)	Brazil (1.36)	**Italy** (1.27)
Kazakhstan (1.16)	Iran (1.35)	**France** (1.19)
South Korea (1.14)	South Africa (1.22)	Iran (1.17)
Spain (1.03)	**Spain** (1.21)	**Poland** (1.01)
Brazil (0.99)	Indonesia (1.16)	South Africa (1.00)
Iran (0.93)	**Poland** (1.14)	**Ukraine** (0.92)
Top 20 countries: 80%	Top 20 countries: 77%	Top 20 countries: 78 %
Contributions by remaining 167 countries: 20% (each country contributed between 0.01% and 0.9%)	Contributions by remaining 167 countries: 23% (each country contributed between 0.01% and 0.9%)	Contributions by remaining 167 countries: 22% (each country contributed between 0.01% and 0.9%)

*   Bold indicates Annex I countries.

Source: Prepared by the first author based on original data from EDGAR tables (Olivier *et al.* 2014).

Table 3.2   Geography of Carbon Dioxide Emissions in 1990: Data for Countries with Low Emission Rates

1990: Total global emissions: 22 Gt Percent of global emissions by each country	Number of countries	Geographical distribution (by regions and countries)
0.1–0.99%	60	Europe: 23 (Annex I countries: 20) Central Asia: 6 Africa and Middle East: 5 Asia and Middle East: 18 South America: 6 Caribbean: 1 (Cuba) Australia and New Zealand: 1 (New Zealand)
0.01–0.09%	66	Europe: 11 (Annex I countries: 6) Africa: 19 Asia and Middle East: 14 South America: 11 Caribbean and Pacific Island nations: 11
Less than 0.01%	47	Europe: 2 (Greenland, Gibraltar: non- Alex I countries) Africa: 26 Asia: 6 (Bhutan, Cambodia, Laos, Maldives, Nepal, Timore-Leste) South America: 2 (Guyana, French Guiana) Caribbean and Pacific Island nations: 11

Source: Prepared by the first author based on original data from EDGAR tables (Olivier *et al.* 2014)

Table 3.3   Carbon Dioxide Emission Reduction Challenges for Annex I Countries, 1990-2000

Failure (emissions in 2000 increased above the 1990 levels)	Success (emissions in 2000 decreased below the 1990 levels)
Cyprus (+57) (i.e., 57% increase above the 1990 level)	Lithuania (-67) (i.e., 67% decrease below the 1990 level)
New Zealand (+52)	Latvia (-64)
Turkey (+51)	Estonia (-60)
Portugal (+49)	Ukraine (-54)
Spain (+35)	Romania (-49)
Ireland (+33)	Belarus (-45)
Australia (+30)	Bulgaria (-43)
Canada (+23)	Russia (-32)
Greece (+23)	Slovakia (-30)
Iceland (+21)	Luxembourg (-25)
USA (+18)	Hungary (-22)
Norway (+14)	Czech Republic (-17)
Slovenia (+13)	Germany (-15)
Japan (+10)	Poland (-7)
Italy (+8)	UK (-7)
Austria (+7)	Malta (-6)
Belgium (+7)	Switzerland (-1)
Netherlands (+7)	Denmark (-1)
France (+4)	
Sweden (+1)	
Finland (+1)	

Source: Prepared by the first author based on original data from EDGAR tables (Olivier *et al.* 2014)

Table 3.4 Geography of Carbon Dioxide Emissions: Per Capita Emissions in 2013

2013 Total global emissions: 35 Gt	2013 Population in millions	2013 Per capita emission: metric tons per person
China (29% of global total)	1,343	7.65
*USA (15)	314	16.88
India (5.87)	1,205	1.72
Russia (5.11)	142	12.65
Japan (3.86)	127	10.68
Germany (2.4)	81	10.39
South Korea (1.78)	49	12.83
Canada (1.56)	34	16.07
Brazil (1.45)	199	2.57
Indonesia (1.38)	249	1.96
Top 10 countries: 68%	3,344 (3.34 billion)	
Saudi Arabia (1.36)	27	18
UK (1.35)	63	7.64
Mexico (1.35)	115	4.13
Iran (1.15)	79	5.17
Australia (1.12)	22	17.93
Italy (1.10)	61	6.36
France (1.04)	66	5.61
Turkey (0.94)	79	4.14
South Africa (0.93)	49	6.76
Poland (0.92)	38	8.42
Top 20 countries: 79%	4,343 (4.34 billion)	

*   Bold indicates Annex I countries.

Source: Prepared by the first author based on original data from EDGAR tables (Olivier *et al.* 2014)

Chapter 4

WHAT IS GLOBAL WARMING?

Definitions and related issues

Global warming is all about temperature increases. It refers to the idea that the Earth's atmosphere has been warming for several decades. Recently, we have learned from television and newspaper reports that the year 2014 was the warmest year in the last 135 years (1880-2014). This may be a new record but in recent years we have become accustomed to similar reports. For example, earlier we were told that the average global temperatures in 2005 and 2010 had broken previous records. According to NASA, the year 2014 also marked the 38th consecutive year since 1977 that the annual (yearly) global temperature was above average. All of these reports indicate that we are living through global warming. The term global warming seems to be a simple expression as it implies increases in global temperatures, but estimating such increases is not an easy task. The most basic problem lies in an inadequate coverage of temperature monitoring stations throughout different parts of continents and oceans. Although oceans cover nearly three-quarters (71%) of the earth surface, data on sea surface temperatures (SSTs) are available from a limited number of monitoring buoys and recording ships. Recently, advances have been made in measuring SSTs by satellite probes. Despite this advancement in technology, there is a major gap in coverage of monitored temperature data from oceans. Even on the continents all areas do not have meteorological stations. Remote areas, such as the Arctic and large deserts like the Sahara, have limited stations deployed. Major mountainous areas also lack observation stations. How do we handle such gaps in global temperature? The next section describes the steps necessary for addressing these issues.

Methods of estimating average global temperatures

Temperature interpolation

The term interpolation means estimating values for the unknowns from known values in the same range. Here is a simple example of temperature interpolation. Suppose we are dealing with three consecutive stations from left to right (or from west to east). Starting from the first station at your left (west), let us suppose it has a temperature of 20°C (68°F). The second station has no data (a blank). The third station has a temperature of 22°C (72°F). We can guess (infer) from such an arrangement (distribution of data) that temperature is increasing towards the east (from left to right). Therefore, it is reasonable to assume that the second station has a temperature of 21°C (70°C). This method of filling temperature data for stations without any data is called temperature interpolation. This is one of the basic principles of estimating temperatures of many places throughout the world which do not have recorded data.

Temperature anomaly

The application of the principle of temperature interpolation using actual station temperature data faces another major hurdle. It is related to changes in air temperature within short distances. Field data indicate that air temperature recorded at a monitoring station may be significantly different from actual temperatures at short distances from that station. In contrast, temperature anomalies seem to remain unchanged for long distances, up to about 1,000 km (621 miles). Temperature anomaly may be defined as the deviation (difference) of a station temperature from its long-term average, say a 30-year average. The advantage of using temperature anomaly data, compared to actual (absolute) data, should be clear. Actual station data may not be interpolated successfully to distant areas with gaps in temperature record, temperature anomaly data can fill up the gap up to relatively long distances from monitoring stations. For this reason, the average global temperatures are estimated by using temperature anomaly data. The application of temperature anomaly data for estimating the average global temperatures involves yet another additional step. This step relates to the expression of temperature anomalies as index values which are neutral of actual temperature units, such as °C or °F. Since such index values do not use specific temperature units these are often called non-dimensional units. The use of relatively easy conversion factors allows one to convert these anomaly index values into absolute temperatures, either in °C or in °F.

International temperature grids

Finally, to obtain global temperature anomalies, the following steps are taken:

- Divide the earth surface into a grid, consisting of a number of boxes (cells) of latitudes and longitudes.
- A decision is made about using a temperature baseline period for the entire Earth for calculating temperature anomalies from this baseline. For example, this could be a 100-year average or a 30-year average temperature of the Earth.
- Next, each grid is filled with temperature anomaly data, which may range from positive (+) anomaly (increase in temperature compared to the long-term average), negative (-) anomaly (decrease in Temperature), or no change.
- Using a computer program, the global average temperature anomalies could then be computed as monthly anomalies or annual (yearly) anomalies.

From the preceding steps it should be clear that estimating average global temperature anomalies (which are then converted into average global temperatures) is actually quite involving and complex. Since the earth's surface is vast, it is also an expensive operation requiring extensive computation processes (computer programs, computer time and manpower resources). The following are the leading international organizations (among others) which have adopted the preceding steps for estimating average global temperatures:

- USA: GISS Surface Temperature Analysis, http://data.giss.nasa.gov/airtemp/ . . .

This is a program by the Goddard Institute for Space Studies (GISS) which is administered by NASA (National Aeronautics and Space Administration). It is perhaps the most comprehensive program for surface temperature analysis as it uses a 2° x 2° latitude/longitude grid system. The GISS uses a 30-year base period (1951-1980) for estimating temperature anomalies.

- USA: NCDC Global Surface Temperature Anomalies, www.ncdc.noaa.gov/ . . .

This is a program by the National Climatic Data Center (NCDC) which is administered by NOAA (National Oceanic and Atmospheric Administration). It uses a 5° x 5° latitude/longitude grid system. The NCDC uses a 100-year base period (1901-2000) for estimating temperature anomalies.

- UK Met Office Hadley Centre HadCRUT4, www.cru.uea.uk/cru/data/temperature/ . . .

This is a joint program of the UK Met Office in collaboration with the University of East Anglia Climate Research Unit (HadCRUT4). It uses a 5° x 5° latitude/longitude grid system. The HadCRUT4 uses a 30-year base period (1961-1990) for estimating temperature anomalies.

Historical trends of global warming

We have started this topic with a simple question: what is global warming? Surprisingly, we had to go through a series of lengthy steps, as described above, just to explain how the average global temperatures are estimated. If the original question is altered slightly, some of the answers have the potential to be even more demanding. One such probing question could be: is global warming really taking place (i.e., is the Earth really warming up)? Or, is it warming everywhere? Or, is it warming at a uniform rate? These are major geographical and slightly more technical questions regarding global warming. The literature on global warming and climate changes in different parts of the world is vast. Instead of reviewing the existing scholarly literature, which would expand this article beyond its scope, we have taken a popular descriptive approach for presenting a set of readily available digital data in the form of summary tables, to describe current global warming trends and to demonstrate how it varies from one place to another. Based on GISS temperature anomaly data for the last 135 years (1880-2014), Table 4.1 summarizes global warming trends during this period of record. The original GISS data were retrieved from a web-based digital file called LOTI: Land-Ocean Temperature Index, which is a report of Combined Land-Surface Air and Sea-Surface Water Temperature Anomalies from 1880 to 2014. As this acronym implies, land-surface temperature anomalies are based on data from land-based monitoring stations (largely weather stations), whereas sea-surface water temperature anomalies are based on SST data from monitoring buoys and ships. The following are some of the major findings from the summary data (Table 4.1).

Long-term average temperatures: The GISS baseline temperature for 1951-1980 was 14 °C (57.2°F). The 135-year average was also 14 °C (57.2°F).

Global warming trends by decades: Temperature changes by decades indicate that the ten-year average temperatures had been increasing gradually since 1921-1930 and it has been increasing above the long-term average of 14 °C (57.2°F) for the last four decades (1971-2010).

Above-average annual temperatures: Annual temperature data (not reported in Table 4.1) indicated that it had been increasing above 14°C (57.2°F) each year since 1977, with a record-breaking 14.68°C (58.42°F) in 2014. Using annual temperature data, it was further evident that the annual temperatures for 59 years out of 78 years since 1937 registered above-average values.

Recent acceleration of warming: The data provide further evidence of recent acceleration of warming, especially in the last two decades of the 20th century and the first decade of this century. Assessed in many different ways, the average global temperature has been rising steadily for nearly a century, especially for the last half a century.

Land and ocean temperatures: In our analyses we have not used separate data for continents and oceans mainly because of the limitation of space. However, at the minimum, it is necessary to stress here that much of the global warming owes its origin to ocean temperatures because water has a much higher heat capacity than land (i.e. heat absorption and retention capacity). For the same reason temperature anomaly data indicate that lands have been significantly cooler than water during the period of analysis. The average SST of oceans during the 20th century (1901-2000) was 16.1 °C (61°F), compared to only 8.5°C (47.3°F) for land areas.

No question about global warming: Despite apparently small increases in average global temperatures, there is no question that global warming is real and that we are living through it. These relatively small increases in average global temperatures have major implications for climate change. The long-term averages tend to hide large increases in temperatures at certain parts of the world and in certain years.

Samples of geographical variations in global warming

Although the preceding summary data provide convincing evidence of global warming, for a casual observer these data may raise questions about the magnitudes (or degrees) of warming. For example, the average temperature of 8.5°C (47.3°F) for the land surface during the 20th century seems to be a relatively cold temperature. Even an average sea surface temperature (SST) of 16.1°C (61°F) does not seem to be a particularly warm temperature. Why is there so much concern about global warming? There are at least two related answers to such a question. First, these data represent global averages of a wide range of temperatures from all temperature grids. Second, such average data represent significant geographical variations in air temperatures of continental areas. For the ease of reference, such variations in temperatures from one place to another may be characterized as *thermal climates*, i.e., climates of different places based only on their air temperatures. The thermal climate of a country or any region of a continent is controlled by a combination of geographical factors. Among them, the basic factors include: latitude (for example, hot tropical climates vs. cool mid-latitude climates), altitude (cooler climates of mountainous areas due to high elevation), and distance from oceans (coastal marine climates vs. interior continental climates). Because of such geographical variations in thermal climates all areas of continents have not warmed up at a uniform rate. Some areas have become even cooler than others. Assessing global warming in different parts of the world is a challenging proposition. This is especially difficult because of differences in the rates of warming in different types of climates. We have chosen the following sample of three countries somewhat arbitrarily, just to demonstrate how recent global warming rates have varied in three different climates with their contrasting temperature and precipitation regimes.

- Canada: a mid-latitude cold and snow climate
- Bangladesh: a tropical hot and rainy climate
- Saudi Arabia: a tropical and subtropical hot and desert climate

Canada

Justifications for selecting three stations

Canada is a vast country. With an area of 10 million square kilometer (3.86 million miles), it is the second largest country in the world (following Russia). It has an extensive network of weather stations, many with long-term temperature and precipitation records, which are readily available from the Environment Canada websites. We have chosen only three weather stations somewhat arbitrarily because of the limitation of space, but with an assumption that these three stations represent fairly well the dominant climates of Canada from the Atlantic to the Pacific coasts. The justifications for choosing these stations are described below at some length.

General characteristics of Canadian climates

Overall, Canada has a cold climate with significant amounts of snowfall in the winter. There are two basic reasons for its cold climate:

(a) its location in higher latitudes (ranging from mid-latitudes to subarctic, arctic, and polar climatic zones) and (b) its vast size for which the bulk of the country is situated far from moderating influences of oceans. Because of the lack of oceanic influences most of the climatic types of Canada may be characterized as continental (i.e., belonging to the interior of a continent). In such a climate, annual temperature ranges tend to be extreme, that is, summer temperatures may be hot to very hot and winter temperatures are likely to be bitterly cold. Average January temperatures in many stations are below 0°C (32°F). Because of cold winter temperatures, winter precipitation in most of the stations occur as snowfall and the average annual temperatures of most of the Canadian stations are below 10°C (50°F).

Winnipeg

According to a more formal (textbook) climatic classification, most of the Canadian population live in two subtypes of continental climates, called *continental climate with a cool summer* and *continental climate with a warm summer*. The first type occurs in a crescentic belt extending from the Atlantic coast (Maritime Provinces) to the Rocky Mountains at the Alberta/British Columbia border. This is by far the largest climatic zone of southern settled part of Canada. Some of the typical stations of this climate include Montreal, Ottawa, Winnipeg, Edmonton, Calgary, Saskatoon, Regina (and many more cities throughout Canada). Out of these potential choices, Winnipeg represents the colder version of this climate fairly well as its annual average temperature (2.52°C or 36.5°F) is significantly lower than that of most of the other stations of this climate.

Toronto

Toronto experiences the second version of the continental climate, that is, with a warm summer. It represents a relatively narrow band of land in southern Ontario through the Great Lakes which extends southward into the northern parts of the United States. Like other stations of southern Ontario, Toronto has much warmer temperatures than Winnipeg throughout the year, as indicated by its average annual temperature of 7.74°C (46°F) but, like Winnipeg, it also experiences large temperature ranges, often with very hot summers and cool to cold winters.

Vancouver

The only climate of Canada that differs significantly from continental climates is a type of *marine climate* that occurs in southwestern British Columbia along the Pacific coast. Defined in terms of the average annual temperature, this is the warmest climate zone of Canada with annual mean temperatures of most of the stations exceeding 10°C (50°F). It differs from other Canadian climates in another notable respect, that is, most of its precipitation occurs as rainfall, largely throughout the winter months. In other words, this is the only climatic type of Canada that is not normally characterized as snow climate. As most of its rainfall occurs during the winter months, the summer tends to be dry under the influence of rapidly shifting high-pressure systems (cells). Since this type of precipitation regime (i.e., rainy winters and dry summers) resembles the Mediterranean climate, in some of the textbooks this type of marine climate has been characterized as *coastal Mediterranean climate*. Vancouver is a typical station of this climate which occurs mainly along a relatively narrow coastal belt of British Columbia.

Global warming trends in sample stations

The preceding review may seem to be a bit long as a preamble to the main topic, but considering its vast size and its numerous climatic types we have made an attempt to provide some contexts for exploring global warming trends in southern settled parts of Canada. The data from Table 4.2 provide the following warming trends during a period of seven decades (1941-2010).

Overall warming trend: As in the case of historical trends of average global temperatures (Table 4.1), each of the three Canadian stations has registered warming trends during the period of observation.

Variations in warming trends among sample stations: Toronto has registered the highest rates of increases in average annual temperatures from 7.5°C (45.5°F) in 1941-1950 to 8.84°C (48°F) in 2001-2010. For the comparable period, this increase in Toronto's average temperature (by 1.34°C or by 2.5°F in six decades) is significantly higher than Vancouver's increase of 0.71°C (by 1.27°F) and Winnipeg's 0.60°C (by 1°F). Averaged for a period of three decades, the mean annual temperature of Toronto has increased from 7.27°C (45°F) in 1951-1980 to 8.17°C (46.7°F) in 2001-2010 (an increase of 0.9°C or 1.62°F), compared to increases of 0.79°C or 1.42°F in Winnipeg and 0.61 °C or 1°F in Vancouver.

Warmest years: There was no predictable pattern for the record warmest years as both Toronto and Vancouver experienced their highest annual temperatures (11.5°C or 52.7°F) in 1954, whereas Winnipeg's highest was in 1987 (5.4°C or 41.72°F). More recently, in 2004, both Toronto and Vancouver recorded their second highest annual temperature of 11.4 °C (52.52°F).

Recent acceleration of warming: The data also provide evidence of greater rates of warming in more recent years. For example, out of 70 years of record (1941-2010), Toronto's annual temperatures were above-average in 31 years. Two-thirds of this record (21 years) were established during the last three decades (1981-2010); the remaining one-third (10 years) occurred in the previous four decades (1941-1981). Vancouver recorded above-average temperatures in 24 years during the same period (1981-2010), compared to Winnipeg's 16 years (Table 4.2).

Thus, assessed in many different ways, data in Table 4.2 confirm the overall trends of Table 4.1 indicating that the three sample stations of Canada have experienced steady increases in their average annual temperatures during the last seven decades (period of observation), especially for the last three decades.

Bangladesh

Climate change hotspot

Bangladesh is a hotspot for global warming-induced climate change. Historical data indicate that frequencies of floods and high-magnitude cyclones (hurricanes) in Bangladesh have increased during the last six decades. Sea levels have been rising along the Bay of Bengal coast of Bangladesh at least for the last four decades at rates of about 4 mm/year (0.16 inches/year). Most of the climate change models project (predict) that by the end of this century about one-fifth of Bangladesh is likely to be inundated due to sea level rise. Climate scientists have implicated global warming for such changes because it affects ocean water temperatures directly. A warmer ocean, in turn, not only expands its volume (leading to sea level rise) but also tends to produce intense low-pressure systems (over oceans) some of which may convert into tropical cyclones (hurricanes). Bangladesh is now ranked as the *most vulnerable country* (MVC) for climate change impacts. Again, climate scientists have implicated greenhouse gas emissions by industrial nations for such adverse climate changes in Bangladesh and other third world countries. Like China, USA and India, Canada is also a major contributor to global greenhouse gas emissions (at least by per capita emission). In contrast, Bangladesh contributes very little to global greenhouse gas emissions (less than 0.1% of global total). Bangladesh seems to be an ideal case study for global warming in a third world country, especially in the context of global warming-causing greenhouse gases originating from distant industrial nations.

General characteristics of Bangladesh climate

Bangladesh has a tropical monsoon climate with a hot and rainy summer and a very short and dry winter (mostly in December and January) with slightly lower temperatures than in other months. Because of relatively warm temperatures throughout the year the annual range of temperature is low. For example, the difference between the warmest month temperature (27°C or 80°F in May and June) and the coldest month temperature (18°C or 64.4°F in January) is only 9°C (difference of 15.6°F) during the 110-year period of analysis, i.e., 1900-2009. There are hardly any geographical variations in air temperature throughout the country, owing to its small size (about one-fifth of Saskatchewan and one-fifteenth of Saudi Arabia) and its location within the warm/hot tropical latitudes (Table 4.3). The selection of three sample stations is based on minor north-south temperature differences.

Global warming trends in sample stations

The data from Table 4.3, based on data from the World Bank Group: Climate Change Knowledge Portal, provide the following warming trends during a period of 110 years (1900-2009).

Overall warming trends: The average annual temperature of Bangladesh during the period of analysis (1900-2009) was 24.78°C. Compared to this long-term average, its average annual temperature in 1990-2009 was 25.30 °C, an increase of 0.52 °C (Table 4.3). This may seem to be slightly lower than the global average increase of 0.68°C for a comparable period. However, based on the World Bank aggregate data, which shows temperature changes by increments of three decades (and the last two decades), warming rates have accelerated in the later part of the record. Thus, the average annual temperature in 1990-2009 has increased by 1.28°C or by 2.31°F (from 24.02 °C or 75.23°F in 1900-1930 to 25.30°C or 77.54°F in 1990-2009). Further, these increases by successive decades have been consistently above the 110-year average (24.78 °C or76.6°F) since the 1930s (Table 4.3), implying warming trends throughout the last seven decades of the 20th century and the first decade of the 21st century.

Regional variations in temperature changes: Among the three stations, temperature changes have been less consistent for Dinajpur, situated in northwestern Bangladesh. Compared to the national trend, it has registered a decline in its average annual temperature in 1990-2009 to 22.61°C (72.7°F), compared to 24.64 °C (76.3°F) in 1900-1930 (a decline of 2.03 °C or 3.6°F). Dhaka, the capital of Bangladesh, situated at the centre of the country, has registered relatively small increases of 0.26°C to 0.31°C (increases of 0.4°F to 0.55°F). Increases have also been less consistent for Chittagong, a coastal city situated in southeastern Bangladesh, registering increases of 0.42°F to 0.43 °C (0.75°F to 0.77°F) in 1930-1960 and 1990-2009, respectively but a slightly lower increase in 1960-1990 (an increase by 0.24 °C or by 0.43°F).

Seasonal warming trends: When temperature data were compared by individual months (not shown in Table 4.3), increases in seasonal temperatures were within the long-term ranges. Thus, winter temperatures (December and January) increased by 0.9°C to 1.3°C (by 1.62°F to 2.34°F). Similarly, summer temperatures (May-August) increased by 1.0°C to 1.4°C (by 1.8°F to 2.52°F). In other words, temperatures have increased in both winters and summers without significant differences in rates of warming between the seasons.

Saudi Arabia

Origin of hot and hyper-arid climate

About one-half of Saudi Arabia is located in tropical latitudes (south of 23.5° N, i.e., the Tropic of Cancer); the rest is in subtropical latitudes (between 25° N and 32° N). Despite its partial location in tropical latitudes why does Saudi Arabia have a hyper-arid (desert) climate with less than 250 mm (10 inches) of annual rainfall? The main reason is the influence of the subtropical high pressure system. In a low-pressure system, moist air rise upward leading to cooling, condensation, cloud formation and precipitation. In contrast, in a high-pressure system, air sinks downward from higher levels of the atmosphere, producing two significant physical changes in the air. First, sinking air (technically called subsidence) compresses air, resulting in increases in air pressure. Second, compression results in warming of the air. Pumping a bicycle tire may be a helpful analogy for understanding this process. When we pump a bicycle tire, not only the air pressure inside the tire increases but the tire also becomes warmer. Something like this happens in the subtropical high-pressure systems. Sinking air increases both air pressure and air temperature by compression. As the air sinks downward it also suppresses convection and rainfall. That is why, all major deserts of the world lie under the influence of the subtropical high pressure systems (most of the time of the year).

Continental climate: With an area of 2.15 million square kilometer, Saudi Arabia is one of the largest countries in the world (ranking 14th by size). Like Canada, large areas of the country are situated far from oceanic influences and consequently many stations experience continental climates or at least climates with significant continental tendencies. Because of this effect most of the stations experience hot summers and cooler winters (much hotter and cooler than those of Bangladesh).

Effect of altitudes: High elevations in northern plateaus and the Assir Mountain ranges in southwestern Saudi Arabia, about 100 km (62 miles) inland of the Red Sea coast, have significant cooling effects on most of the stations in these regions.

Monsoon rainfall: The southwestern region of Saudi Arabia along the Red Sea coast, including the Assir Mountain, is also the only region of the country that receives significant amounts of precipitation from incursions of summer monsoons from the Arabian Sea. The annual precipitation in this region is about 430-435 mm (about 17 inches), far exceeding the limit of 250 mm (10 inches), which is the boundary of the desert climate. Because of higher elevations some of this precipitation may occur even as snowfall and sleet in this region.

Sample stations: Clearly, the climate of Saudi Arabia is not as homogenous as we might have implied initially by our characterization of it as a hot and hyper-arid climate. On further analyses, our choice of three sample stations represents our assumptions of differences in regional climates of Saudi Arabia:

- Abha, situated in the Assir Mountain region, represents the influence of high elevations.
- Riyadh, the capital of Saudi Arabia and situated at the interior of the country, represents continental influence.
- Dhahran, situated on the coast of the Arabian/Persian Gulf, represents marine influence.

Global warming trends in sample stations

The data from Table 4.4 provide the following warming trends during the period of analysis (1900-2009).

Overall warming trend: The average annual temperature of Saudi Arabia during the last two decades of analysis (1990-2009) was 25.3°C (77.54°F), compared to its long-term average of 24.45°C (76°F) (1900-2009). The resulting increase by 0.85°C (by 1.53°F) for the comparable period was higher than both the global average increase by 0.68°C (by 1.22°F) and an average increase by 0.52°C (by 0.93°F) for Bangladesh. As in the case of Bangladesh, increments by periods of three decades (and the last two decades) demonstrate similar acceleration of warming in the later part of observation. Thus, the average annual temperature in 1990-2009 has increased by 1.78°C (from 23.52°C in 1900-1930 to 25.30°C in 1990-2009) or by 3.2°F (from 74.3°F in 1900-1930 to 77.54°F in 2009). As in the case of Bangladesh, again, these increases by successive decades have been consistently above the long-term average (25.30°C or 77.54°F) since the 1930s (Table 4.4), implying warming trends throughout the last seven decades of the 20th century and the first decade of the 21st century.

Regional variations in temperature changes: The warming trend in sample stations, measured as increments by periods of three decades between 1900 and 1990, was less consistent for Abha and Dhahran than that for Riyadh. Whereas, Riyadh registered increases in average annual temperatures during each of the two periods following the base period (1900-1930), Abha registered slight declines in both of these periods. Like Riyadh, Dhahran registered slight increases in both of these periods but its average temperature in the last two decades of analysis (26.38°C or 79.48°F in 1990-2009) registered a decline, compared to its long-term average of 26.75°C (80.15°F). Despite its situation on the coast of the Arabian/Persian Gulf, its average temperatures do not indicate any moderating effect of a marine climate. However, the average annual temperatures at both Riyadh and Abha have increased significantly during the last two decades of analysis (1990-2009), indicating recent acceleration of warming.

Seasonal warming trends: The data on average annual temperatures are presented in Table 4.4 by increments of multiple decades. They seem to be remarkably similar to those of Bangladesh, with the exception of the data for Abha, a high-altitude station. This apparent similarity is misleading for Saudi Arabia because its average annual temperature data hide significant seasonal differences in temperatures. The data in Table 4.5, comparing increases in seasonal temperatures, provide a much better picture of the magnitudes of warming in Saudi Arabia. These data show that the rates of increases in seasonal temperatures averaged for entire Saudi Arabia seem to be within its ranges of long-term increases. Thus, winter temperatures for Saudi Arabia have increased by 0.6°C to 0.8°C (by 1°F to 1.44°F), whereas summer temperatures have increased by 1.1°C to 1.2°C (by 2°F to 2.16°F) (Table 4.5). In contrast, much greater rates of warming have occurred at Abha: its winter temperatures have increased by 1.27°C to 2.36°C (by 2.29°F to 4.25°F), whereas its summer temperatures have increased by much higher rates, by 2.51°C to 2.98°C (by 4.52°F to 5.36°F) (Table 4.5). All of the data in this table confirm that the summers in Saudi Arabia have become warmer. This may seem to be contrary to our expectation of warmer winter temperatures due to global warming, but in this case warmer summers may be attributed to the shifting seasonal positions of the subtropical high pressure system. This would certainly require additional advanced research on this topic.

Limitations of this study and concluding comments

The findings of our research should be treated with caution at least for three reasons. **First**, our findings are contingent upon the quality of the original data we have used in this study. We have used the GISS data on global temperature anomalies because these are now widely used for estimating global warming trends. These data may further be verified by comparing them with similar other data sources, such as the NCDC data by NOAAA and the UK Hadley Centre data (HadCRUT4). The long-term temperature data for Canadian stations, obtained from the Environment Canada websites, could be verified readily. Our findings on global warming trends in Bangladesh and Saudi Arabia are based on digital data on air temperatures for these countries (among many others), which have been posted by the World Bank Group under its Climate Change Knowledge Portal. Regarding reliability of these data, the World Bank web portal (data source) states that its data are "quality controlled". In particular, it states that the historical temperature data have been provided by the Global Historical Climatology Network. **Second**, our selection of sample countries and stations has been somewhat arbitrary and, therefore, we do not claim that our findings are adequately representative of geographical variations in global warming. Yet, the main justification of our sample selection was to demonstrate how the rates of recent global warming have varied from one area to another, especially from one climate to another. **Third**, our method of analysis is essentially descriptive. Although we have used large amounts of data we have presented them as summary tables; we have not used any statistical analyses, such as the *difference of means* tests or *trend analysis* of long-term data. Consistent with the central goal of this book, we have tried to use a popular discourse (language of communication) for presenting scientific and technical concepts. Notwithstanding these limitations, we can make the following generalizations safely based on our findings.

- First, conforming to popular perceptions (or at least perceptions by the majority), there is no question that global warming is real and that we are living through it. The data also provide evidence of recent acceleration of warming, especially during the last two decades of the 20th century and the first decade of this century.
- Second, as we assumed, rates of global warming have varied from one region to another, especially from one climate to another. The average annual temperature of Toronto, representing a cold and snow climate, has increased at a higher rate than that of Vancouver, a warm and rainy marine climate. As in the case of the general global warming trend, all of the three Canadian stations provide evidence of recent acceleration of warming.
- Third, all of the sample stations in Bangladesh have registered increases in their average annual temperatures, again with acceleration of warming in recent years.
- Fourth, contrary to our initial assumption, the rates of warming have been higher in the hot and hyper-arid (desert) climate of Saudi Arabia than in the hot and rainy (tropical monsoon) climate of Bangladesh. Similarly, contrary to our expectation, summers have become warmer in Saudi Arabia than its winters. Further research is needed to explain the causes of such changes, especially to test our assumption that such temperature anomalies could be attributed to seasonal shifts in the subtropical high pressure system.

References

Besides original data sources acknowledged below each of the tables, we have consulted the following sources for preparing this chapter.

Barry, R.G. and Chorley, R.J. 2010. *Atmosphere, Weather and Climate* (9th Edition). New York and London: Routledge.

Houghton, J. 2004. *Global Warming: The Complete Briefing* (3rd Edition). Cambridge, UK, New York: Cambridge University Press.

Rashid, H. and Paul, B. 2014. *Climate Change in Bangladesh: Confronting Impending Disasters*. Lanham, MD, Boulder, CO, New York, Toronto, Plymouth, UK: Lexington Books.

Table 4.1 Historical Trends of Average Global Temperatures, 1881-2014

Trend by decade:	Ten-year average temperature (°C)	Ten-year average temperature (°F)
1881-1890	13.78	56.80
1891-1900	13.75	56.75
1901-1910	13.64	56.55
1911-1920	13.67	56.61
1921-1930	13.80	56.84
1931-1940	13.94	57.09
1941-1950	13.98	57.16
1951-1960	13.97	57.15
1961-1970	13.99	57.18
1971-1980	14.04	57.27
1981-1990	14.22	57.60
1991-2000	14.36	57.85
2001-2010	14.58	58.24
1881-2014 (135 average)	14.0	57.2
1977-2014 (each year): >14.0	>14	>57.2
1937-2014: 59 years out of 78 years)	Above-average temperatures	Above average temperatures
1881-1937: 57 years	Below-average temperatures	Below-average temperatures

Source: Prepared by the first author based on annual temperature anomaly data provided by NASA's GISS Surface Temperature Analysis (see GISS end note for the method of converting anomaly data into absolute temperatures). Temperature anomaly data by GISS are available at: http://data.giss.nasa.gov/gistemp/station_data/ . . . retrieved in January 2015.

Table 4.2   Historical Trends of Average Annual Temperatures at Toronto, Winnipeg, and Vancouver, Canada, 1941-2010

Sample characteristics	Toronto 43°42' N 79° 46' W	Winnipeg 49° 53' N 97° 10' W	Vancouver 49° 13' N 123° 06' W
*Average annual temperature in °C (°F)	7.74 (45.93)	2.52 (36.54)	10.04 (50.07)
Ten-year averages:			
1941-1950	7.5 (45.5)	2.28 (36.10)	9.8 (49.64)
1951-1960	7.98 (46.36)	2.68 (36.82)	9.89 (49.80)
1961-1970	7 (44.60)	1.95 (36.51)	9.86 (49.75)
1971-1980	6.82 (44.28)	1.97 (36.55)	9.6 (49.28)
1981-1990	7.57 (45.63)	3.15 (37.67)	10.13 (50.23)
1991-2000	8.09 (46.56)	2.72 (36.90)	10.53 (50.95)
2001-2010**	8.84 (47.91)	3.18 (37.72)	10.51 (50.92)
Thirty year averages:			
1951-1981	7.27 (45.09)	2.2 (35.96)	9.78 (49.60)
1981-2010***	8.17 (46.71)	2.99 (37.38)	10.39 (50.70)
Number of years with above-average annual temperature (1981-2010)	21	16	24
Number of years with above-average annual temperature (1941-1980)	10	15	10

*   Average annual temperature is calculated as the average of twelve monthly temperatures (January to December). Monthly temperatures are the averages of daily temperatures.

**   For Winnipeg: 2001-2006

*** For Winnipeg: 1981-2006

Source: Prepared by the first author based on original historical monthly data from Environment Canada. http://climate.weather.gc.ca/ClimateData/monthlydata . . . retrieved on 8 February 2015

Table 4.3 Historical Trends of Average Annual Temperatures at Dinajpur, Dhaka, and Chittagong, Bangladesh, 1900-2009

Bangladesh and its sample stations→	Bangladesh	Dinajpur 25° 38' N 88° 44' E	Dhaka 23° 42' N 90° 22' E	Chittagong 22° 20' N 91° 48' E
Periods	Average annual temperature in °C (°F)	Average annual temperature in °C (°F)	Average annual temperature in °C (°F)	Average annual temperature in °C (°F)
1900-2009	24.78 (76.60)	24.46 (76.03)	25.64 (78.15)	25.19 (77.34)
1900-1930	24.02 (75.24)	24.64 (76.35)	25.40 (77.72)	25.04 (77.07)
1930-1960	24.98 (76.96)	24.98 (76.96)	25.71 (78.28)	25.46 (77.83)
1960-1990	25.07 (77.13)	24.97 (76.95)	25.70 (78.26)	25.28 (77.50)
1990-2009	25.30 (77.54)	22.61 (72.70)	25.66 (78.19)	25.47 (77.85)

Source: Prepared by the first author based on original historical monthly temperature data compiled by the World Bank Group: Climate Change Knowledge Portal, available at: http://sdwebx.worldbank.org/climateportal/ . . . retrieved on 18 February 2015

Table 4.4   Historical Trends of Average Annual Temperatures at Abha, Riyadh, and Dhahran, Saudi Arabia, 1900-2009

Saudi Arabia and its sample stations→	Saudi Arabia	Abha 18° 14' N 42° 31' E	Riyadh 24° 39' N 46° 46' E	Dhahran 26° 13' N 50° 02' E
Periods	Average annual temperature in °C (°F)	Average annual temperature in °C (°F)	Average annual temperature in °C (°F)	Average annual temperature in °C (°F)
1900-2009	**24.45 (76.01)**	**16.93 (62.47)**	**24.69 (76.44)**	**26.75 (80.15)**
1900-1930	23.52 (74.34)	16.63 (61.93)	24.42 (75.96)	26.71 (80.08)
1930-1960	24.63 (76.33)	16.41 (61.54)	25 (77)	26.77 (80.19)
1960-1990	24.65 (76.37)	16.51 (61.72)	24.60 (76.28)	27.10 (80.78)
1990-2009	25.30 (77.54)	18.83 (65.89)	25.82 (78.48)	26.38 (79.48)

Source: Prepared by the first author based on original historical monthly temperature data compiled by the World Bank Group: Climate Change Knowledge Portal, available at: http://sdwebx.worldbank.org/climateportal/ . . . retrieved on 18 February 2015

Table 4.5   Seasonal Increases in Air Temperature in Saudi Arabia, 1900-2009

Saudi Arabia	1900-1930 in °C (°F)	1990-2009 °C (°F)	Increases (+ °C) (+ °F)
Winter:			
January	14.7 (58.46)	15.3 (59.54)	0.6 (1.08)
February	16.2 (61.16)	17 (62.6)	0.8 (1.44)
December	16 (60.8)	16.8 (62.24)	0.8 (1.44)
Summer:			
June	30 (86)	31.2 (88.16)	1.2 (2.16)
July	30.5 (86.9)	31.6 (88.88)	1.1 (1.98)
August	30.7 (87.26)	31.8 (89.24)	1.1 (1.98)
Abha	1900-1930 in °C (°F)	1990-2009 °C (°F)	Increases (+ °C) (+ °F)
Winter:			
January	10.81 (51.46)	12.08 (53.74)	1.27 (2.28)
February	11.68 (53.02)	13.69 (56.64)	2.01 (3.62)
December	11.56 (52.81)	13.92 (57.06)	2.36 (4.25)
Summer:			
June	21.47 (70.65)	24.09 (75.36)	2.62 (4.71)
July	21.83 (71.29)	24.34 (75.81)	2.51 (4.52)
August	21.9 (71.42)	24.88 (76.78)	2.98 (5.36)

Source: Prepared by the first author based on original historical monthly temperature data compiled by the World Bank Group: Climate Change Knowledge Portal, available at: http://sdwebx.worldbank.org/climateportal/ . . . retrieved on 18 February 2015

Chapter 5

WHAT IS CLIMATE CHANGE?

Three distinct components of climate change

In less than thirty years, *climate change* has evolved into an exceedingly complex topic because of the way it is used by different interest groups. The latter include diverse groups, among them:

- Scientists researching and writing on climate change
- Policy makers: politicians responsible for legislations on climate change and bureaucrats responsible for implementing these policies
- Industrialists and businessmen who have an interest in minimizing expenditures on emission control (i.e., control of carbon dioxide emissions responsible for global warming)
- Environmentalists and activists who have an interest in protecting the environment from the adverse impacts of climate change
- News media reporting on different aspects of climate change and climate change policies

One of the central problems with public discourse (language of communication) on climate change is that each of the preceding groups may focus on entirely different aspects of climate change, using different interpretations of the topic. Even many scientists writing on different components of global warming, climate change and climate change impacts may label all of these topics under the general name of climate change. We believe that much of the current controversies associated with climate change could be minimized by distinguishing these topics into three categories.

First, *global warming* refers to increases in the average global temperatures. This is an objective term as the average temperatures of the Earth can be verified by recorded data. **Second**, as the term *climate change* stands for literally, it deals with noticeable changes in climatic patterns of different places on the earth surface. This is also an objective concept as many of the changes in climatic patterns can be verified by evidences (climatic data and field evidences of changes), though not necessarily conclusively. **Third**, *climate change impacts* are less objective as these are subject to contesting interpretations. A cause-and-effect relationship between global warming or climate change and its impact on a specific characteristic of the atmosphere is difficult to prove. It requires lots of sophisticated research. Recent changes/increases in the frequencies of cyclones/hurricanes provide an example. Advanced research on this topic have tried to link cyclone activities in the Pacific and the Indian Oceans to changes in air pressure over warmer ocean waters during El-Niño events in the Pacific and the Indian Ocean Dipole (IOD) events. Similarly, research has established links between summer monsoon floods in India and Bangladesh to the La Nina events over the Pacific Ocean—a cooler Pacific east coast but a warmer west coast. In short, proving climate change impacts is a much more challenging proposition than either providing objective data on global warming or describing changes in climatic patterns of a given place/region.

Nature of this topic

Climate change can be studied at different scales. At a global scale, some regions have already experienced major changes in their climatic patterns. For example, droughts have expanded over large parts of the Sahel in West Africa. Lately, heat waves have become common in some parts of Europe and the amounts of rainfall have increased in certain parts of the Tropics. Based on extensive data analyses, many scientific articles have been published on such regional patterns of climate change. Because of the limitation of space, we have taken a much narrower approach for studying changes in local climatic patterns. Thus, in this study we have analyzed long-term air temperature data for Toronto and precipitation data for Vancouver, to assess if such data provide any evidence of changes in local climatic patterns of these two contrasting geographical locations.

To test our assumptions of changes in local climatic patterns, we have asked several familiar questions. Has the climate of Toronto become warmer in recent years? Similarly, have summers become warmer and winters milder in Toronto? Toronto has a continental climate because of its location in the interior of North America. Therefore, one of our basic assumptions was that such an interior place might be an ideal location for studying changes in its thermal climatic patterns. In contrast, Vancouver is situated on the west coast of Canada on the Pacific Ocean which enjoys a moderate climate because of the influence of the ocean. Therefore, changes in air temperatures are not a major concern here. Instead, it has often been claimed anecdotally that Vancouver has become rainier in recent years. Some have even suggested that the precipitation patterns of Vancouver have changed, implying changes in seasonal concentrations of summer and winter precipitation. We have summarized long-term data in the form of descriptive tables to answer some of these basic questions.

Has the climate of Toronto become warmer in recent years?

The answer is yes! To answer this question we have analyzed long-term temperature data in two separate steps.

Normal air temperatures

At the first step, we have compared *normal* air temperatures for Toronto for a period of 50 years (1961-2010). In climate science, the term normal means long-term average. Following the World Meteorological Organization (WMO) practice, Environment Canada uses a 30-year average for calculating *Canadian Climatic Normals*. The

Environment Canada website has posted three successive Canadian Climatic Normals: (a) 1961-1990, (b) 1971-2000, and (c) 1981-2010. The latest normal data for 1981-2010 are based on ten-year updates of the previous three decades of data (1971-2000). Similarly, the latter was an update of the previous three decades of data (1961-1990). Data are provided for both air temperature and precipitation, using different types of statistics. The simplest of these statistics may be expressed as *normal annual temperature*, which is the average annual temperature for 30 years. The *average annual temperature*, in turn, is the average of daily temperatures for 365 days of each year. According to these Canadian Climatic Normal data, the normal annual temperatures for Toronto were as follows:

- 1961-1990: 7.2°C (45°F)
- 1971-2000: 7.5°C (45.5°F)
- 1981-2010: 8.2°C (46.76°F)

These data show that the average annual temperature of Toronto has increased by 1°C (by 1.8°F) during the last 50 years (1961-2010). For the equivalent period the global average temperature has increased by about 0.59°C (by 1.06°F). Thus, Toronto has experienced greater rates of warming than the global average. This finding is consistent with our expectation as continental places are likely to experience greater rates of warming than the global average, which includes both ocean surface and land surface temperatures and data from wide ranges of latitudes. As a note of caution, this is one of our basic assumptions that would still require further data analyses to confirm it. Because of the space limitation we have not made any attempt to compare data from different continental and marine locations.

Average temperatures by decades

At the second step, we have further expanded our analysis by comparing temperature increases by successive decades during a period of 70 years (1941-2010) (Table 5.1). These data show that warming has been gradual and more consistent for the last three decades (1981-2010), compared to the previous four decades (1941-1980). Whereas the first two decades (1941-1960) enjoyed warmer than the 1961-1990 normal temperature of 7.2°C (45°F), the third and the fourth decades (1961-1980) experienced cooler than normal temperatures. Starting from 1981, Toronto's annual temperatures have been increasing steadily. Another way to look at the data in Table 5.1 is to compare the latest normal temperature of 8.2°C (46.76°F) with the entire data set. As the last column in the table indicates, average annual temperatures were cooler than the latest normal for the first six decades (1941-2000). Only the last decade (2001-2010) has registered higher than normal annual temperature. The largest amount of increase has occurred in this decade, registering an increase of +0.64°C (1.15°F) above the 1981-2010 normal. The data for 2011 and 2012 indicate not only continuation of warming trends but also abnormally large increases of annual temperatures, by as much as +0.9°C (1.62°F) and +2.3°C (4.14°F), respectively, above the 30-year normal.

Have summers become warmer in Toronto?

The answer is again yes! The following are our analyses.

Summer temperature trends

Although we have not employed any advanced statistical analyses, we have conducted extensive recalculations of different types of data to answer this question. Some of these data require definitions as they are slightly technical. Two such definitions are as follows:

- The *daily temperature* of a given place (a weather station, for example) is the average of hourly temperatures for 24 hours of a day.
- The *normal daily temperature* of a month refers to the latest 30-year average daily temperature for that month (averaged for 30-31 days of the month).

Using daily data reports for a period of 70 years (1941-2010), we have calculated 10-year averages of daily temperatures for the months of May, June, July, August and September (five warmest months) to determine summer temperature trends by successive decades, compared to the latest 30-year normal daily temperatures (1981-2010) (Table 5.2).

For example, the normal daily temperature for the month of May (1981-2010) is 13.10°C (55.58°F). The 10-year average data for May indicate that for five decades (1941-1990) the average daily temperatures were cooler than the 1981-2010 normal. It was only in the last two decades, starting from 1991, that the daily temperatures for the month of May have increased by +0.28°C (by 0.54°F) in 1991-2000 and by +0.17 °C (by 0.3°F) in 2001-2010. Overall, the 10-year data show considerable variations in the detailed patterns of temperature changes in the summer months. Some of the trends are as follows:

- The latest 10-year averages (2001-2010) for all of the five months have increased compared to the first decade of analysis (1941-1950).
- None of the months has registered consecutive increases in average daily temperatures for three decades in a row. Out of the five months, the average daily temperatures have increased for two decades in a row (1991-2010) only in the months of May and June. This implies that increases in the latest normal daily temperatures (1981-2010) have been contributed largely by above-normal increases in the last decade (2001-2010).
- The greatest amount of increase has occurred in the month of September, from 15.72°C (60.3°F) in 1941-1950 to 17.56°C (63.6°F) in 2001-2010 (an increase of +1.84 °C or 3.31°F).
- The average temperature in the month of July, the hottest month, has increased from 20.76°C (69.37°F) in 1941-1950 to 22.05°C (71.69°F) in 2001-2010 (an increase of 1.29°C or 2.32°F).

Daily maximum temperatures in summer months

Using data for the daily maximum temperatures (one of the items of the Canadian climatic normal data), we have also compared how the 30-year averages of daily maximum temperatures in a given month have changed during the last 50 years (1961-2010) (Table 5.3). The original data on *daily maximum temperatures* in a given month are the 30-year averages of 30-31 daily maximum temperatures. Calculated in this manner, changes in the daily maximum temperatures have been more consistent than the general summer temperature trends (data presented in Table 5.2). Data in Table 5.3 show that the average daily maximums for each of the five months have increased gradually from 1961-1990 to 1981-2010. Again, the largest increase has occurred in the month of September, from 20.9°C (69.62°F) in 1961-1990 to 21.6 °C (70.88°F) in 1981-2010 (an increase of +0.7 °C or 1.26°F).

In the last row of Table 5.3 we have listed five *highest daily temperatures*, one for each of the five months, during a period of 70 years (1941-2010). Each of these was the absolute highest daily temperature (record temperature) in a given month in 70 years. From the dates of these records, it seems that they had occurred in random, independently from recent warming trends. Four of the five records occurred during the earlier years (1948-1962). The highest daily temperature in 70 years was 38.3°C (100.94°F), which was recorded on 25 August 1948. The second highest temperature (37.6°C or 99.68°F) was recorded on 7 July 1988. It is relevant here that, irrespective of their timing, each of the record temperatures represented heat alert events. For the city of Toronto, *heat alerts* are issued when the daily maximum temperatures reach a threshold value of 30°C. When such high temperatures (30°C or 86°F or higher) continue consecutively for a period of three to seven days, these events are characterized as *heat waves*. Some studies suggest that heat alerts and heat waves have increased in Toronto in recent years.

Have winters become milder in Toronto?

The answer is yes! To answer this question, we have prepared Table 5.4, which is identical to the format/nature of Table 5.2. Data on 10-year averages of daily temperatures for the months of December, January and February (three coldest months) indicate patterns of changes in winter temperatures, which have some similarities with summer temperature trends. Thus:

- Daily temperatures have become significantly warmer in the months of December, January and February in the last two decades (1991-2010), compared to the first decade of analysis (1941-1950).
- The average temperature for the month of January, the coldest month, has increased by 1.25°C (by 2.25°F), from -6.04°C

(21.13°F) in 1941-1950 to -4.79°C (23.38°F) in 2001-2010. This also represented an increase of +0.71°C (1.28°F) from the latest normal temperature for January (-5.50 °C or 22.1°F)).

- The average temperatures in the third and the fourth decades (1961-1980) were particularly cooler than their long-term normal trends. In contrast, the last decade (2001-2010), in particular, has experienced milder winter temperatures.

Has Vancouver become rainier in recent years?

The answer is yes! To answer this question, first, we have analyzed 70 years of data on annual precipitation by successive decades (Table 5.5).

Annual precipitation trends in Vancouver

The following trends are evident from changes in the amounts of annual precipitation by decades during the period of analysis:

- Annual precipitation has increased in every decade from 1941-1950 to 2001-2010. The absolute increase in 70 years has been by an amount of +100 mm or about 4 inches (from 1027 mm or 40.41 inches to 1127 mm or 44.31 inches).
- The latest normal (1981-2010) annual precipitation for Vancouver (1189 mm or 46.81 inches) has been exaggerated by abnormally high precipitation in 1981-1990 (1227 mm or 48.30 inches). Consequently, gradual increases in annual precipitation by successive decades before and after this record period seem to be somewhat masked by this spike in rainfall. With a random decline in 2011, the data for 2012 indicate continuation of long-term increases in precipitation.
- The seasonal concentration of precipitation (mostly rainfall) does not indicate any particular pattern changes. Roughly, two-thirds to three-quarters of rainfall occur during the wettest half of the year (October to March).
- The data on snowfall confirm that Vancouver has a predominantly rainy climate as snowfall amounts are insignificant compared to the annual rainfall amounts, especially if snowfall is converted

into equivalent rainfall (roughly by a factor of 1 cm of snowfall is equal to about 1 mm of rainfall).

Winter precipitation trends in Vancouver

We have prepared Table 5.6 to assess if winter precipitation has increased in Vancouver. In Table 5.5, earlier we looked into seasonal concentration of precipitation during the wettest half of the year. In Table 5.6, we have analyzed 50-year trends (1961-2010) in winter precipitation in the months of November, December and January (three rainiest months of the year). These data show that the successive 30-year averages (normal data) for winter precipitation have increased in the months of November and January, whereas December has registered a decline. We are unable to explain this decline in precipitation in the month of December without further research.

In the last row of Table 5.6 we have listed data for three highest daily rainfall and snowfall, one for each of the three months, during a period of 50 years (1961-2010). Each of these was the absolute highest daily precipitation (record precipitation) in a given month in 50 years. As in the case of record temperatures in Table 5.3, it seems that they occurred in random, independently from recent increases in annual precipitation. The highest daily rainfall in 50 years was 89.4 mm (3.52 inches), which was recorded on 25 December 1972. The second highest amount was 68.3 mm (2.69 inches), which occurred on 18 January 1968. The third highest amount on 3 November 1989 was 65mm (2.55 inches), which was not significantly different from the second highest amount. In short, there is no conclusive evidence suggesting that the seasonal precipitation patterns in Vancouver have changed significantly.

Conclusion

In this chapter we have analyzed long-term temperature data for Toronto and precipitation data for Vancouver to assess if global warming has caused any significant changes in local climatic patterns of these two contrasting places (stations). Although our selection of sample stations has been arbitrary we were constrained by space limitation and by our overall goal of a popular style of analysis. We did not wish to conduct any sophisticated statistical analyses to assess changes in local climatic patterns. Instead, our data analyses have been elementary, involving recalculations of averages of original data and determining differences of these averages from long-term normal data (30-year averages). Despite such a basic approach, we found that our data analyses have provided some evidences of recent changes in air temperature patterns (thermal climates) of Toronto and precipitation patterns of Vancouver. It should be stressed here that for explaining why or how these changes have occurred, more advanced research would be required employing different types of climatic/meteorological data and sophisticated statistical analyses.

Consistent with the style of the popular science, we have tried to answer some of the familiar questions. Two of our major findings are as follows:

- Our analyses indicate that there is no question that global warming has affected the thermal climatic patterns of Toronto. Overall, Toronto has become warmer in recent years. Its summers have also become warmer and winters have become milder.
- Global warming seems to have affected the precipitation patterns of Vancouver. As it has been claimed anecdotally, Vancouver has indeed become rainier during the last 70 years. However, our data analyses do not provide evidence of any significant changes in the seasonal concentrations of rainfall in Vancouver.

References

Besides original data sources acknowledged below each of the tables, we have consulted the following sources for preparing this chapter.

Barry, R.G. and Chorley, R.J. 2010. *Atmosphere, Weather and Climate* (9th Edition). New York and London: Routledge.

Houghton, J. 2004. *Global Warming: The Complete Briefing* (3rd Edition). Cambridge, UK, New York: Cambridge University Press.

Rashid, H. and Paul, B. 2014. *Climate Change in Bangladesh: Confronting Impending Disasters*. Lanham, MD, Boulder, CO, New York, Toronto, Plymouth, UK: Lexington Books.

Table 5.1   Annual Temperature Trends in Toronto, 1941-2012

Period	Annual temperature in °C (°F)	Higher than normal annual temperature by °C (by °F)	Lower than normal annual temperature by °C (by °F)
1941-1950	7.5 (45.5)		0.7 (1.26)
1951-1960	7.98 (46.36)		0.22 (0.4)
1961-1970	7 (44.6)		1.2 (2.16)
1971-1980	6.82 (44.28)		1.38 (2.48)
1981-1990	7.57 (45.62)		0.65 (1.17)
1991-2000	8.09 (46.56)		0.11 (0.2)
2001-2010	8.84 (47.91)	0.64 (1.15)	
2011	9.1 (48.38)	0.9 (1.62)	
2012	10.5 (50.9)	2.3 (4.14)	

*   Normal annual temperature for Toronto is 8.2°C (46.76°F), which is the average of the annual temperatures for the latest 30 years (1981-2010).

Source: Prepared by the first author based on original data from Environment Canada, Historical Climate Data, Monthly Data Reports (1941-2012) and Canadian Climatic Normals (1981-2010). http://www. climate.weather.gc.ca/climateData/ . . . retrieved on 2 March 2015.

Table 5.2 Summer Temperature Trends in Toronto, 1941-2012: (a) 10-year average daily temperature in °C/°F, (b) higher than (+) or lower than (-) *normal daily temperature by °C (°F)

Month	1941-1951	1951-1960	1961-1970	1971-1980
May	(a) 12.32/54.2 (b) -0.78 (-1.4)	(a) 12.58/54.64 (b) -0.52 (-0.94)	(a) 11.75/53.15 (b) -1.35 (-2.43)	(a) 12.44/54.39 (b) -0.66 (-1.19)
June	(a) 18.43/65.2 (b) -0.17 (-0.3)	(a) 18.52/65.34 (b) -0.02 (-0.04)	(a) 17.56 /63.6 (b) -1.04 (-1.87)	(a) 17.1/62.78 (b) -1.5 (-2.7)
July	(a) 20.76 /69.37 (b) -0.74 (-1.33)	(a) 21.36/70.45 (b) -0.14 (-0.25)	(a) 20.11/68.2 (b) -1.39 (-2.5)	(a) 20.19/68.34 (b) -1.31 (-2.35)
August	(a) 20.31/68.56 (b) -0.20 (-0.52)	(a) 20.52/68.94 (b) -0.08 (-0.14)	(a) 19.05/66.29 (b) -1.55 (-2.79)	(a) 19.5/67.1 (b) -1.1 (-1.98)
September	(a) 15.72/60.3 (b) -0.48 (-0.86)	(a) 16.18/61.12 (b) -0.02 (-0.04)	(a) 15.3/59.54 (b) -0.9 (-1.62)	(a) 14.9/58.82 (b) -1.3 (-2.34)
Month	**1981-1990**	**1991-2000**	**2001-2010**	
May	(a) 12.71/54.88 (b) -0.39 (-0.7)	(a) 13.38/56.08 (b) +0.28 (+0.5)	(a) 13.27/55.89 (b) (b) +0.17 (+0.3)	
June	(a) 17.55/63.59 (b) -1.05 (-1.89)	(a) 18.84/65.91 (b) +0.24 (+0.43)	(a) 19.42 /66.96 (b) +0.82 (+1.48)	
July	(a) 21.2/70.16 (b) -0.3 (-0.54)	(a) 21.09//70 (b) +0.41 (+0.74)	(a) 22.05/71.69 (b) +0.55 (+0.99)	
August	(a) 19.77/67.59 (b) -0.83 (-1.49)	(a) 20.29/68.52 (b) -0.31 (-0.56)	(a) 21.61/70.9 (b) +1.01 (+1.81)	
September	(a) 15.34/59.61 (b) -0.86 (-1.54)	(a) 15.72/60.3 (b) (b)-0.48 (-0.86)	(a) 17.56/63.6 (b) +1.36 (+2.44)	

* Normal daily temperature refers to the latest 30-year (1981-2010) average daily temperature for a given month. The normal daily temperatures for 1981-2010 were as follows: May: 13.10°C (55.58°F), June: 18.60°C (65.48), July: 21.50°C (70.7°F), August: 20.60°C (69.08°F), September: 16.20°C (61.16°F).

Source: Prepared by the first author based on original data from Environment Canada, Historical Climate Data, Daily Data Reports (1937-2012) and Canadian Climate Normals (1961-2010). http://www.climate.weather.gc.ca/climateData/ . . . retrieved on 2 March 2015.

Table 5.3 Daily Maximum Temperatures in Summer Months in Toronto, 1961-2010

Period	May °C (°F)	June °C (°F)	July °C (°F)	August °C (°F)	September °C (°F)
1961-1990 (30-year average)	18.4 (65.12)	23.6 (74.48)	26.8 (80.24)	25.5 (77.9)	20.9 (69.62)
1971-2000 (30-year average)	18.8 (65.84)	23.7 (74.66)	26.8 (80.24)	25.6 (78.08)	21 (69.8)
1981-2010 (30-year average)	18.8 (65.84)	24.2 (75.76)	27.1 (80.78)	26 (78.8)	21.6 (70.88)
Historical record	May °C/°F	June °C/°F	July °C/°F	August °C/°F	September °C/°F
1941-1970 (70-year highest temperature with date)	34.4 (93.92) 16 May 1962	36.7 (98.06) 25 June 1952	37.6 (99.68) 7 July 1988	38.3 (100.94) 25 August 1948	36.7 (98.06) 2 September 1953

Source: Prepared by the first author based on original data from Environment Canada, Historical Climate Data, Canadian Climate Normals (1961-2010). http://www.climate.weather.gc.ca/climateData/ . . . retrieved on 2 March 2015.

Table 5.4   Winter Temperature Trends in Toronto, 1941-2010: (a) 10-year average daily temperature in °C (°F), (b) cooler or warmer than *normal daily temperature by °C (°F)

Decade	December (a) 10-year average (b) Cooler/warmer than normal	January (a) 10-year average (b) Cooler/warmer than normal	February (a) 10-year average (b) Cooler/warmer than normal
1941-1950	-3.76°C (25.23°F): cooler by 1.56 °C (by 2.8°F)	-6.04°C (21.13°F): cooler by 0.54°C (0.97°F)	-6.32°C (20.62°F): cooler by 1.82°C (3.28°F)
1951-1960	-2.84°C (26.88°F): cooler by -0.64°C (by 1.5°F)	-5.73°C (21.69°F): cooler by 0.23°C (0.41°F)	-4.59°C (23.74°F): cooler by 0.09°C (0.16°F))
1961-1970	-3.99°C (24.84°F): cooler by 1.79°C (3.22°F)	-7.28°C (18.9°F): cooler by 1.78°C (3.2°F)	-6.58°C (20.16°F): cooler by -2.08°C (3.74°F)
1971-1980	-3.84°C (25°F): cooler by 1.64°C (2.95°F)	-7.2°C (19°F): cooler by 1.7°C (3°F)	-7°C (19.4°F): cooler by 2.5°C (4.5°F)
1981-1990	-2.88°C (26.88°F): cooler by 0.68°C (1.22°F)	-6.1°C (21°F): cooler by 0.6°C (1.08°F)	-4.88°C (23.21°F): cooler by 0.38°C (0.68°F)
1991-2000	-2.04°C (28.32°F): warmer by 0.16°C (0.29°F)	-5.67°C (21.8°F): cooler by 0.17°C (0.3°F)	-4.43°C (24): warmer by 0.07°C (0.13°F)
2001-2010	-1.64°C (29°F): warmer by 0.56°C (1°F)	-4.79 C (23.38°F): warmer by 0.71C (1.28°F)	-4.35°C (24.17°F): warmer by 0.15°C (0.27°F)

*   Normal daily temperature refers to the latest 30-year (1981-2010) average daily temperature for a given month. Normal daily temperatures for 1981-2010 were as follows: December: -2.2 °C (35.96°F), January: -5.5 °C (41.9°F), February: -4.5 °C (40.1°F).

Source: Prepared by the first author based on original data from Environment Canada, Historical Climate Data, Daily Data Reports (1937-2012) and Canadian Climate Normals (1961-2010). http://www. climate.weather.gc.ca/climateData/ . . . retrieved on 8nd March 2015.

Table 5.5   Annual Precipitation Trends in Vancouver, 1941-2012

	Annual precipitation in mm (inches) Higher or lower than the *normal annual precipitation: +/-%	Annual snowfall in cm (inches)	October-March precipitation in mm (inches): % of annual total
1941-1950	1027 (40.41): -13.62%	42 (16.41)	735 (28.89): 72%
1951-1960	1064 (41.88): -10.51%	53 (20.96)	790 (31.1): 74%
1961-1970	1113 (43.82): -6.39%	62 (24.4)	821 (32.32): 74%
1971-1980	1161 (45.7): -2.35%	66 (26)	747(29.4): 64%
1981-1990	1227 (48.3): +3.2%	37 (14.5)	879 (34.57):72%
1991-2000	1112 (43.77): -6.48%	40 (15.82)	846 (33.26): 70%
2001-2010	1127 (44.31): -5.21%	17 (6.69)	834 (32.77): 74%
2011	1069 (42.08): -10.09%	24 (9.44)	711 (27.99): 67%
2012	1211 (47.67): +1.85%	27 (10.62)	963 (37.91): 80%

*   Normal annual precipitation amounts for Vancouver are as follows: 1981-2010: 1189 mm (46.81 inches), 1971-2000: 1199 mm (47.16 inches), 1961-1990: 1157.4 mm (45.56 inches). These are the averages of the annual precipitation for 30 years (defined by Environment Canada as its normal values). For our analysis of changes from normal in the second column (indicated by +/- sign) we have used the latest (1981-2010) normal (which is 1189 mm or 46.81 inches).

Source: Prepared by the first author based on original data from Environment Canada, Historical Climate Data, Daily Data Reports (1937-2012) and Canadian Climate Normals (1961-2010). http://www. climate.weather.gc.ca/climateData/ . . . retrieved on 8nd March 2015.

Table 5.6 Winter Precipitation Trends in Vancouver, 1961-2010

Period	November Monthly rainfall in mm (inches) [Snowfall in cm (inches)]	December Monthly rainfall in mm (inches) [Snowfall in cm (inches)]	January Monthly rainfall in mm (inches) [Snowfall in cm (inches)]
1961-1990 (30-year average)	167.2 (6.58) [2.6 (1.02)]	161.2 (6.34) [18.6 (7.32)]	131.6 (5.18) [20.6 (8.11)]
1971-2000 (30-year average)	178.5 (7) [2.5 (0.98)]	160.6 (6.32) [16.3 (6.42)]	139.1 (5.48) [16.6 (6.53)]
1981-2010 (30-year average)	185.8 (7.31) [3.2 (1.26)]	148.3 (58.3) [14.8 (5.82)]	157.5 (6.2) [11.1 (4.37)]
Historical record	Rainfall in mm (inches): Date [Snowfall in cm (inches): Date:]	Rainfall in mm (inches): Date [Snowfall in cm (inches): Date:]	Rainfall in mm (inches): Date [Snowfall in cm (inches): Date:]
Highest daily rainfall [snowfall] in 50 years (1961-2010)	65 mm (2.55): 3 November 1989 [22.1 cm (8.7): 30 November 1975]	89.4 mm (3.52): 25 December 1972 [31.2 cm (12.28): 31 December 1968]	68.3 mm (2.69): 18 January 1968 [29.7 cm (11.69): 13 January 1971

Source: Prepared by the first author based on original data from Environment Canada, Historical Climate Data, Canadian Climate Normals (1961-2010). http://www.climate.weather.gc.ca/climateData/ . . . retrieved on 2 March 2015.

Chapter 6

CLIMATE CHANGE IMPACTS ON BAY OF BENGAL CYCLONES

Basic problems

It is much harder to prove climate change impacts on cyclones or floods than determining global warming trends from recorded data. Historical data on cyclones/hurricanes provide some clues to their frequencies and magnitudes in the recent past, but they do not necessarily provide a straight-forward case of cause-and-effect relationships between global warming and cyclone frequencies or magnitudes. Since cyclones/hurricanes originate over oceans, potential relationships between global warming and cyclonic activities may be based on at least two related assumptions. **First**, oceans are likely to absorb increasing quantities of heat from the atmosphere due to global warming, resulting in increases in sea surface temperatures (SSTs). **Second**, increased SSTs are expected to result in increased evaporation, precipitation, formation of low pressures and stronger winds, i.e. increased cyclonic activities. One of the problems with SSTs is that they are not necessarily uniform throughout an entire ocean body. Instead, they tend to develop anomalies, called SST anomalies, in the form of large pools of abnormally warm or abnormally cold water in certain parts of the ocean. The air temperature, air pressure and cyclonic activities of the atmosphere immediately above the sea surface are affected by the temperature characteristics of these pools. Since atmosphere is an open system warmer or cooler air associated with these SST anomalies may not necessarily remain confined within the boundaries of a given ocean. The general circulation of the atmosphere (major wind systems) may transport these warmer or cooler air masses from one ocean to another or from one area to another. Thus, cyclones over a given ocean may be affected not only by its own SST anomalies but also by such anomalies in other distant oceans. In short, cyclones are impacted by a chain of inter-related factors: (a) general warming of the atmosphere (i.e., global warming), (b) impact of global warming on SSTs, (c) mutual exchange of energy (feedbacks) between SSTs and the atmosphere, and (d) general circulation of the atmosphere redistributing impacts of SST anomalies from different oceans. Using historical data on cyclone frequencies and magnitudes in the Bay of Bengal and timing of SST anomalies in the Indian Ocean and the Pacific Ocean, we have made an attempt to isolate and analyze the main issues of these complex problems. Thus, the main objectives of this article are as follows:

First, analyze historical data on frequencies and magnitudes of cyclones in the Bay of Bengal to determine their trends during a period of more than one century.

Second, review characteristics of temperature anomalies in the Indian Ocean (called the Indian Ocean Dipole or IOD) and the Pacific Ocean (called El Niño and La Niña), to assess how these might be inter-related despite great distances between them.

Third, review the latest data on the numbers of Bay of Bengal cyclones which occurred simultaneously with SST anomalies in both the Indian Ocean and the Pacific Ocean.

Popular assumptions about climate change impacts on cyclones

Two major assumptions in public discourse regarding climate change impacts on tropical cyclones in Bangladesh can be summed up as follows:

- First, it has been claimed repeatedly by many scientists, environmentalists, politicians and many people living in coastal Bangladesh that the Bay of Bengal cyclones have become more frequent and more violent in recent years.
- Second, it has also been suggested that the assumed acceleration of cyclonic activities have probably been the results of climate change.

To verify these two assumptions, we have analyzed two types of data: (a) long-term historical data on frequencies and magnitudes (categories) of tropical cyclones (hurricanes) in the Bay of Bengal and (b) assumed linkages between cyclonic activities and periodic changes in positions of warm/cool water pools in the Pacific and Indian Oceans, more specifically changes in relative positions of El Niño, La Niña, and the Indian Ocean Dipole (IOD) events. Assumed increases in cyclonic activities in both of these oceans have been attributed to such warm water pools in tropical oceans.

Have cyclones become more frequent and more violent in the Bay of Bengal?

The answer is maybe or maybe not. It depends on how one would frame the question. Using data in Tables 6.2 and 6.3, we would like to demonstrate how the answers differ significantly from one another, depending upon the nature of the questions.

Nature of the questions

The language of communication (discourse) is very critical for describing changes in cyclone frequencies and their magnitudes in the Bay of Bengal, especially because we are dealing here with data for two different geographical situations. **First**, data in Table 6.2 refer to all cyclonic activities in the Bay of Bengal. Many of those storms did not strike Bangladesh. Some had landfalls over India and Myanmar but many simply dissipated over the ocean. **Second**, in contrast, when people talk about impacts of cyclones in coastal Bangladesh probably they imply impacts of storms which had struck the coast of Bangladesh.

We will analyze this specific concept under a separate section below by using data in Table 6.3.

Figure 6.1 Collapsed roof of a bamboo-frame house damaged by *Cyclone Gorky*, 1991. Photo courtesy: Emdad Haque, University of Manitoba, reproduced with permission.

Frequencies of cyclonic storms

Dealing with the first question, we have summarised 117 years of data (1891-2007) on the numbers of cyclonic storms that had developed over the Bay of Bengal during this period (Table 6.2). These data include all categories of tropical cyclones (categories 1-5), in addition to *tropical storms* that start from a minimum wind speed of 39 mph (miles per hour) (63 km/h) (Table 6.1). According to the Saffir-Simpson scale, a tropical cyclone has a threshold wind speed of a minimum of 74 mph (120 km/h) (Table 6.1). The maximum speeds include category 5 cyclones (>157 mph or 252 km/h). We would like to stress here that our data in Table 6.2 are secondary in nature, as we have drawn them from a recent publication by Mahala and others (2015), but the original data used by these authors are from a primary source, namely the cyclone e-Atlas of India Meteorological Department.

The very first finding from the data in Table 6.2 seems to be contrary to our expectation, as there was a declining trend in cyclonic activities in the Bay of Bengal for the entire period of record (1891-2007). To verify this finding, let us look at the details of this trend by comparing the number of storms by successive decades. In all, there were 502 storms in 117 years with an average frequency of 4.29 storms per year. Out of the eleven decades (1891-2000), the numbers of storms in six decades were above the long-term average, but they all occurred in the first half of the record (i.e., between 1891 and 1950). The maximum frequency (6 per year) was in 1920-1930, whereas the minimum (2.8 per year) was in 1991-2000. With the exception of 1961-1980, when there were slight increases in frequencies of storms (with above-average values), overall, the declining trend started after 1950 and the latest data (2001-2007) seemed to continue this trend. Thus, we can conclude from these data that, contrary to popular expectation, frequencies of storms/cyclones in the Bay of Bengal have declined over a period of at least one century. It should be stressed here that the bulk of the data presented in Table 6.2 consists of less intense tropical storms and lower categories of tropical cyclones.

Frequencies of high magnitude cyclones

Table 6.3 addresses the question if tropical cyclones striking coastal Bangladesh have become more intense and more violent in recent years. The answer is a definite yes based on a 127 years of storm track information for the landfall of tropical cyclones along the Bangladesh coast. Although our data in Table 6.3 are again secondary in nature as we have obtained them from a publication by Islam and Peterson (2009), the authors of this article have used primary data from the *Global Tropical Cyclone Climatic Atlas* (See endnote 2 for the details of these sources).

In all, there were 115 storms in 127 years (1877-2003). This is a post-Industrial Revolution period during which carbon dioxide had been accumulating in the atmosphere for more than 100 years. For the first 82 years of record (1877-1959) there is no consistent pattern of increases in landfall of cyclones along the Bangladesh coast. However, starting from 1960, there was a striking increase in the landfall of tropical storms and tropical cyclones in coastal Bangladesh. There were 29 tropical storms and 18 tropical cyclones between 1960 and 2003. Out of the 26 tropical cyclones in 127 years, 18 of them, that is, more than two-thirds (69%) occurred between 1960 and 2003. Not only was there a significant spike in cyclone activities during this relatively short period, perhaps more significantly, all of the five super-cyclones (with wind speeds exceeding 138 mph, 222 km/h) occurred during this period, providing evidence of recent intensification of tropical cyclones over northern Bay of Bengal. These storms have also become extremely destructive causing massive property damage and loss of lives. It has been estimated by an International Disaster Database (see endnote 3) that between 1970 and 2012 cyclone storm surges had been responsible for the loss of nearly 500,000 lives in coastal Bangladesh. Further, increasing numbers of coastal residents have been affected by these cyclones in many different ways. Most significantly, millions had been displaced temporarily during some of the catastrophic cyclones. Although there are no reliable data on population displacements due to cyclones, it has been widely reported that large numbers of coastal residents have also been displaced permanently (that is, uprooted from their homes).

Nature of Sea Surface Temperature (SST) anomalies

Teleconnections

Some of the latest research on Bay of Bengal cyclones suggests that recent intensification of these storms may be related to global warming-induced SST anomalies, not only in the Indian Ocean but also in distant Pacific Ocean. The SST anomalies in the Pacific Ocean are known as El Niño and La Niña, whereas their Indian Ocean counterpart is called the Indian Ocean Dipole (IOD). How could these distant phenomena be interrelated over distances of thousands of miles/kilometer? There are at least two potential interconnections. First, the *prevailing winds* (i.e., the dominant winds of the general atmospheric circulation) transport warmer/cooler air from one area to another, since the global atmosphere is an open system. Second, warmer/cooler ocean water might be transported from the Pacific to the Indian Oceans (and vice versa) through openings between Australia and Indonesia. Further, these SST anomalies (El Niño, La Niña, and IOD) are coupled ocean-atmosphere phenomena. The term *coupled ocean-atmosphere phenomena* means that changes in SSTs result in simultaneous changes in air temperature and air pressure over the sea surface. In short, SST, air temperature and air pressure are interrelated. Thus, changes in these characteristics in either the Pacific Ocean or in the Indian Ocean may affect each other's climatic characteristics. Such distant connections of the global atmosphere are called *teleconnections*.

What are El Niño and La Niña?

These are SST anomalies in the Pacific Ocean, which develop in cycles of warmer and cooler SSTs in alternating geographical positions within the equatorial/tropical belt of the Pacific Ocean. To explain these anomalies, it is convenient to consider three distinct phases of SSTs in the Pacific Ocean.

Phase 1: The **normal** or **neutral** phase does not experience either El Niño or La Niña. During this phase, the SST in western Pacific is warmer than in eastern Pacific. The prevailing winds over the tropical Pacific are easterlies, originating over East Pacific and blowing towards West Pacific. As the easterlies blow constantly over the surface water towards the west, this has a "skimming effect", removing a relatively shallow surface layer of warm water towards West Pacific. The lost warm surface water is then replaced by cooler water from below by a process called *upwelling* (similar to bubbling up). This is a very significant process for fish stocks and other marine life in the upper parts of the ocean where life-sustaining nutrients are delivered from deeper water below by upwelling. Because of strong upwelling of cool water along the coasts of Ecuador, Peru and northern Chile, this area enjoys a productive fishing industry during normal periods. Cool surface water also means very little evaporation. Consequently, the entire coastal area of Ecuador, Peru and Chile experiences a dry desert climate.

Phase 2. **El Niño** is the name for the warmer phase of SSTs in East Pacific. By technical definition by the World Meteorological Organization, to qualify as an El Niño condition the SSTs must be warmer than normal by at least $+0.9°F$ or $+0.5°C$ for three consecutive months and lasting for a minimum of five months. Most of the El Niño (and alternating La Niña) episodes within the last 50 years occurred in cycles of 3-5 years. On average, each episode lasted between one and two years. Exceptionally long ones persisted for periods of up to 34 consecutive months with temperature anomalies reaching as much as ± 2°C to 3°C (3.6°F to 5.4°F). During major El Niño episodes the easterlies (east to west airflows) weaken significantly and may even reverse their directions (flowing west to east). This allows accumulation of a large pool of warm water in East Pacific, extending along the equator all the way from Central Pacific (north of Tahiti) to the coasts of Ecuador and Peru. Accumulation of unusually warm water in East Pacific results in thickening of the surface layer of warm water (up to 1000 ft or 300 m, compared to the normal of 130 ft or 40 m). This, in turn, reduces upwelling of cooler water (resulting in lack of nutrient delivery to the surface layer from deeper water below) with devastating effects for the fish stock along the coasts of Ecuador, Peru and Chile. Extensive fish kills and reductions of fish stocks occur during El Niño.

The climatic effects of El Niño have also been devastating for coasts of Ecuador and Peru. Excessive rainfall and stormy conditions often had resulted in flash floods and landslides. The El Niño effects could also be far-reaching beyond the equatorial belt reaching as far north as the subtropical and mid-latitude belts of North America. Whereas winters in northern Mexico and California tend to be wetter during El Niño episodes, both northwestern and northeastern parts of North America experience drier and warmer winters. As expected, cyclonic activities increase over East Pacific. Most of the high-magnitude hurricanes in East Pacific also occur during the El Niño events.

*Phase 3. **La Niña*** is the name for the reverse condition of El Niño. The easterlies become stronger than normal. Strong easterlies drive warmer surface water towards West Pacific accumulating a large pool of unusually warm water northeast of Australia and Papua New Guinea and east of Indonesia. Cooler water develops in East Pacific (cooler at least by -0.9°F or -0.5°C for three consecutive months and lasting for at least five months). During La Niña, rainfalls and cyclonic activities increase in northern and northeastern Australia and in many parts of Southeast Asia. Even high magnitude floods in Bangladesh have been attributed partly to the La Niña effect.

What is ENSO?

In teleconnection research, effects of SSTs on changing air temperatures and air pressures over the Pacific (responsible for its cyclonic activities) are attributed to ENSO (El Niño-Southern Oscillation), a classic example of a single large-scale coupled ocean-atmosphere phenomenon. The Southern Oscillation (SO) is measured as an index called Southern Oscillation Index (SOI). It measures the differences or fluctuations in air pressure between Darwin (northern Australia) and Tahiti (central Pacific). During negative SOI (called low phase) air pressure is high over Darwin and low over Tahiti and East Pacific, resulting in a westerly flow (west to east), that is, a reversal of the easterlies. The resulting accumulation of warmer SSTs in Central and East Pacific leads to El Niño. During positive SOI (high phase) air pressure is high over the Central Pacific (Tahiti) and low over Darwin, resulting in strengthening of the easterlies, promoting La Niña (see also endnote 4 on *Walker Circulation*).

What is IOD?

The Indian Ocean Dipole (IOD) is the Indian Ocean counterpart of the Pacific El Niño and La Niña. The term dipole means two "poles" or two areas of differences. The IOD measures differences in SSTs between the Arabian Sea (western pole) and the eastern Indian Ocean south of Indonesia (eastern pole). Both of these poles are situated within the equatorial belt of the Indian Ocean (i.e., between 10° N and 10° S) but they have a northwest-southeast diagonal orientation because of the physical configuration of the North Indian Ocean. The Arabian Sea is located north of the equator and the Indian Ocean south of Indonesia is located south of the equator.

Like ENSO, IOD is a coupled ocean-atmosphere phenomenon. The shifting pools of warm/cool water contribute to variations in rainfall and storm activities of many countries surrounding the Indian Ocean. It is postulated that the IOD is linked to the Pacific ENSO through the easterlies (considered as a limb/part of the *Walker Circulation*. See endnote 4) and through transport of warm waters from the Pacific into the Indian Ocean. During a positive IOD warmer SSTs develop over western Indian Ocean (Arabian Sea, in particular). During a negative IOD, the opposite happens, that is, the western Indian Ocean becomes cooler with higher air pressures resulting in westerly winds blowing towards the Indian subcontinent (i.e., reversing the prevailing easterlies). Positive IODs are often associated with El Niño and negative IODs with La Niña.

Effects of ENSO and IOD on Bay of Bengal cyclones

Data in Table 6.4 provide evidence of simultaneous occurrence (co-occurrence) of tropical cyclones with ENSO and IOD events in the Bay of Bengal for a period of 117 years (1891-2007). These data are remarkable because they confirm several scientific and popular assumptions. **First**, conforming to the general observation stated above, positive IODs were associated more with El Niño events (42 events that also co-occurred with tropical cyclones) than with La Niña (8 events), whereas 63 La Niña events (and tropical cyclones) co-occurred with negative IOD, compared to only 9 El Niño events (first row of the table). **Second**, the data confirm the assumption of ENSO effects on cyclonic activities in the Bay of Bengal. This is a significant finding because two-thirds of the tropical cyclones (333 out of 502) co-occurred with El Niño and La Niña during three different phases of the IOD. This confirms the significance of teleconnection between SST anomalies in the Pacific and the Indian Oceans. **Third**, the dominant effect of ENSO events on tropical cyclones in the Bay of Bengal can be seen from the data indicating that, compared to 66% of total cyclones during ENSO events, only about 30% (153 cyclones) occurred during IOD events (both positive and negative). **Fourth**, it is remarkable that even in the absence of IOD at least 42% of these cyclones (127 + 84 = 211 out of 502 = 42%) occurred during alternating phases of El Niño and La Niña, again confirming teleconnection. **Finally**, it is equally remarkable that only about one-quarter of the tropical cyclones (138 out of 502 = 27%) occurred in the absence of both ENSO and IOD effects. This confirms the assumption that SST anomalies (ENSO and IOD) play a significant role in producing nearly three-quarters of tropical cyclones in the Bay of Bengal.

Figure 6.2. A cyclone shelter in coastal Bangladesh. Photo courtesy: M. Nawfel Huda and Khairul Matin, reproduced with permission.

Conclusion

We started this topic with a statement that assessing climate change impacts on tropical cyclones in the Bay of Bengal is a complex problem. The scientific literature on this topic is substantial. Because of the space limitation and the scope of this chapter we have been selective in our literature review. Our goal has been to simplify complex issues for a general audience. To do so, we have been able to isolate the main issues as a number of cause-and-effect relationships. The basic problem lies in the SST anomalies. It has been suggested that these anomalies have been exaggerated by global warming as it impacts on complex ocean dynamics. Again, because of the space limitation we have not reviewed the nature of ocean dynamics that are affected significantly by global warming-induced SSTs. These include, among others, ocean currents, vertical distribution of water temperature called *thermocline*, upwelling of cold water, increased or decreased strengths of wind systems (such as easterlies) blowing over the ocean surface, etc. Despite these incomplete explanations, our analyses of historical data on frequencies and magnitudes of tropical cyclones in the Bay of Bengal and parallel data on simultaneous occurrences of ENSO and IOD events indicate the following major findings:

- First, contrary to expectation, frequencies of cyclonic activities in the Bay of Bengal have declined during a period of at least one century.
- Second, conforming to popular expectation, frequencies of high magnitude cyclones that had landfalls over coastal Bangladesh have increased recently. We have not conducted any research to explore how simultaneous decreases in frequencies of cyclonic activities (numbers of low-magnitude storms) in the Bay of Bengal might be related to intensifications of high-magnitude cyclones in northern Bay of Bengal.
- Third, the main connection between global warming and cyclonic activities in the Bay of Bengal seems to be through SST anomalies in both the Indian Ocean (IOD) and the Pacific Ocean (ENSO). These anomalies have become more pronounced (stronger) with continuing global warming.
- Finally historical data indicate that there is a significant connection between the timing of tropical cyclones in the Bay of Bengal and simultaneous occurrences of the Pacific Ocean ENSO and the Indian Ocean IOD, providing evidence of strong regional teleconnection between the climates of these two distant oceans.

We have learned one important lesson from our study that the language of communication (discourse) is a critical issue for minimizing confusion about climate change impacts on Bay of Bengal cyclones. Our findings show that it is not correct to say that climate change has increased cyclonic activities in the Bay of Bengal. Often public/media discourse makes such a generalization. In contrast, we have provided data demonstrating clearly that frequencies of high magnitude cyclones striking coastal Bangladesh have, indeed, increased recently.

Endnotes

1. **Units for measuring air pressure**: Readers may be familiar with the unit **mb** (millibar) for measuring air pressure because it is frequently used in media reports. It is a unit of the old metric system (scientifically called **cgs** units for centimeter gram per second). The modern metric unit (called SI unit for

International System of Units) for air pressure is *pascal*. The conversion from mb into pascal is: 1 mb = 100 pascal. The standard atmospheric pressure is 1013.2 mb. If we multiply 1013.2 mb by 100 it becomes a very long number. To avoid a long number, the metric prefix *hecto* (h for 100) is added to pascal (pa) to derive a new SI unit called *hectopascal*, short-formed as hpa (100 pa). This is a convenient unit because both mb and hpa are readily interchangeable referring to the same amount of pressure. Thus 1013.2 mb = 1013.2 hpa.

2. The data for the Global *Tropical Cyclone Climatic Atlas* are based on a compilation of several sources of primary data. These include the following: (a) two historical digital tape deck data files archived at the US National Climatic Data Center (NCDC), (b) a Joint Typhoon Warning Center (JTWC) historical data file from Guam, and (c) data forwarded to NCDC by several specialized meteorological centres participating in a Tropical Cyclone Program, sponsored by the UN's World Meteorological Organization (WMO) (Islam and Peterson 2009).

3. **International Disaster Database**: This database is maintained by the Centre for Research on the Epidemiology of Disasters (CRED), Université Catholique de Louvain-Brussels, Belgium. It is also short-formed as EM-DAT: The OFDA/CRED International Disaster Database (www.emdat.be).

4. **Walker Circulation**: Air pressure varies widely from one place to another and from one latitude belt to another belt. At a global scale the general circulation of the atmosphere is based on a model of a semi-permanent high pressure system over the subtropical belt (called subtropical high pressure belt) over 25°-35° N (north of the equator) and 25°-35° S (south of the equator). Prevailing winds (dominant winds) blow in a north-south direction (to begin with) along the meridians (longitudes) from the subtropical high pressure belts to the equatorial low pressure belt (between 10° N and 10° S). Earth's rotation (*Coriolis effect*) deflects these winds as northeasterlis (north of the equator) and southeasterlies (south of the equator). Since these winds blow along the meridian (in a north-south direction to begin with) these are also called meridian flows.

In contrast to meridian flows, the *Walker Circulation* (discovered by Sir Gilbert Walker, the Chief Meteorologist of British India in the 1920s) refers to zonal flows that develop as east-west winds over the Pacific. Conceptually, Walker tied these east-west flows to the Southern Oscillation, which is measured as the differences in air pressure between Darwin (northern Australia) and Tahiti (central Pacific). The Walker Circulation reinforces the easterlies (a limb of this zonal flow) during La Niña, whereas during El Niño the Walker Circulation is weakened. For an explanation of how El Nino and Southern Oscillation are coupled as ENSO, see above the section on ENSO.

Another feature of the Walker Circulation is that warm and moist air over the West Pacific rises up (to the top of the troposphere at about 10 miles/16 km above the sea level), exhausts all moistures by condensation and precipitation and becomes cooler, then crosses the Pacific and sinks over East Pacific, strengthening the high pressure system. This reinforces the easterlies. Thus, a complete east-west vertical loop (cell) is created, which is called the Walker Circulation. More recent research has found that multiple Walker Circulation cells extend from the Pacific Ocean to the Indian Ocean to the Atlantic. Thus, the Walker Circulation plays an important role in transporting winds from surfaces of SST anomalies between the Pacific and the Indian Oceans.

References

We have consulted the following sources for preparing this chapter.

Barry, R.G. and Chorley, R.J. 2010. *Atmosphere, Weather and Climate* (9th Edition). New York and London: Routledge.

Bureau of Meteorology, Government of Australia. 2015. "The Indian Ocean Dipole (IOD)", "Walker Circulation". http://www.bom.gov.au . . . retrieved on 22 April 2015.

Houghton, J. 2004. *Global Warming: The Complete Briefing* (3rd Edition). Cambridge, UK, New York: Cambridge University Press.

Islam, T. and Peterson, R.E. 2009. "Climatology of Landfalling Tropical Cyclones in Bangladesh 1877-2003". *Natural Hazards* 48 (1): 115-135.

Mahala, B.K., Nayak, B.K. and Mohanty, P.K. 2015. "Impacts of ENSO and IOD on Tropical Cyclone Activity in the Bay of Bengal'. *Natural Hazards* 75: 1105-1125.

Rashid, H. and Paul, B. 2014. *Climate Change in Bangladesh: Confronting Impending Disasters.* Lanham, MD, Boulder, CO, New York, Toronto, Plymouth, UK: Lexington Books.

Table 6.1 Saffir-Simpson Scale of Tropical Cyclones

Storm category	Wind speed (knots)*	Wind speed (mph)	Wind speed (km/h)	Central low pressure (mb)
*Tropical depression (TD)	33 knots or lower	38 mph or lower	61 km/h or lower	
*Tropical storm (TS)	34-63	39-73	63-117	
Category 1	64-82	74-95	119-153	980 mb or higher
Category 2	83-95	96-110	154-177	965-979
Category 3	96-112	111-129	178-208	945-964
Category 4	113-136	130-156	209-251	920-944
Category 5	137 knots or higher	157 mph or higher	252 km/h or higher	Lower than 920 mb

*   Tropical Depression and Tropical Storm are not a part of the Saffir-Simpson Scale.

1 knot (nautical mile) = 1.15 miles.

Source: Prepared by the first author based on data from NOAA: http://www.nhc.noaa.gov/ . . .

Table 6.2 Numbers of Tropical Cyclones in the Bay of Bengal, 1891-2007

Decade	Numbers of cyclones*	Numbers per decade
1891-1900	50	5
1801-1910	43	4.3
1911-1920	44	4.4
1921-1930	60	6
1931-1940	52	5.2
1941-1950	45	4.5
1951-1960	31	3.1
1961-1970	51	5.1
1971-1980	48	4.8
1981-1990	35	3.5
1991-2000	28	2.8
2001-2007	15	2.14
1891-2007 total	**502**	**4.29**

*   Include tropical storms with wind speeds: 39-73 miles per hour/63-119 kilometer per hour.

Source: Prepared by the first author based on data presented in Figure 3 in: Mahala, B.K., Nayak, B.K. and Mohanty, P.K. 2015. "Impacts of ENSO and IOD on Tropical Cyclone Activity in the Bay of Bengal". *Natural Hazards* 75: 1105-1125.

Table 6.3   Landfall of Tropical Cyclones Over Coastal Bangladesh, 1877-2003

Decade	TD	TS	Category 1	Cat 2	Cat 3	Cat 4	Cat 5
1877-1888	7	3					
1889-1899	2	1	3*				
1900-1909	2	5	1*				
1910-1919	3	2	1*				
1920-1929	5	3	1*				
1930-1939	6	3	1*				
1940-1949	4	2	1*				
1950-1959	5	2					
1960-1969	5	10	2	1		2**	
1970-1979		8	3			1**	
1980-1989		5	1		1		
1990-2003		6	3		1	2 (1**)	1**
1877-2003 total	**39**	**50**	**17**	**1**	**2**	**5**	**1**

*   Exact wind data not available
** Super-cyclones with wind speed exceeding 120 knots (128 mph/ 222 km/h)

Source: Prepared by the first author based on data presented in Tables 4-15 in: Islam, T. and Peterson, R.E. 2009. "Climatology of Landfalling Tropical Cyclones in Bangladesh 1877-2003". *Natural Hazards* 48 (1): 115-135.

Table 6.4 Simultaneous Occurrence of Tropical Cyclones with ENSO and IOD Events in the Bay of Bengal, 1891-2007

Nature of IOD	Number of tropical cyclones co-occurred with El Niño	Number of tropical cyclones co-occurred with neutral ENSO	Number of tropical cyclones co-occurred with La Niña	Total number of tropical cyclones
Negative IOD	9	9	63	81
No IOD	127	138	84	349
Positive IOD	42	22	8	72
Total	178	169	155	502

Source: Table 3 in: Mahala, B.K., Nayak, B.K. and Mohanty, P.K. 2015. "Impacts of ENSO and IOD on Tropical Cyclone Activity in the Bay of Bengal'. *Natural Hazards* 75: 1105-1125.

Chapter 7

HEAT WAVES IN DELHI
AND TORONTO

Nature of this study

Heat waves have been characterized by Environment Canada as the "silent killer" as this hazard is not visible like floods or cyclones. However, their impacts are highly visible. People die from heat waves—from heatstrokes, dehydration and other heat-related diseases. Heat waves are responsible for all kinds of miseries of daily life: sweats, exhaustion and fatigue, challenges of working outside, difficulties of commuting in hot weather, and so on. To avoid excessive heat, people try to stay indoors, use airconditioning systems excessively (if available), take refuge in airconditioned shopping malls, crowd in swimming pools or in lakeshore beaches (such as in Toronto) or even in nearby rivers (such as in Delhi). The power system is overloaded due to excessive demands for electricity. Excessive water consumption may drain water supply systems. During the May 2015 heat waves in Delhi (including New Delhi, the national capital of India) city streets (black tops) had melted in many blocks, schools had been closed, fewer people had gone to work in offices and factories, and at least 2000 people lost their lives due to heat-related diseases and complications. Over the years, national and international newspapers have published a wide range of reports on different aspects of heat waves in Delhi and Toronto. Almost invariably data on record temperatures and heat wave fatalities have been one of the routine features of such reports. To verify newspaper information on record temperatures, extensive data are available in official websites of the Indian Meteorology Department (IMD) and Environment Canada (EC). In this study, we have used both official data and newspaper reports on record temperatures for studying recent and past heat waves in Delhi, India and Toronto, Canada. We were motivated particularly by the the 2015 May heat waves in Delhi which had attracted widespread attention of the global media.

Besides comparing hard data on air temperature and fatality, there are several other more practical and compelling reasons for using newspapers as a primary source of information on heat waves. **First**, while official websites provide a reliable source of data on air temperatures, specific information on heat wave events are almost impossible to obtain from IMD websites (for Delhi and other Indian stations) and would require further interpretations of EC historical data on *extreme maximum daily temperatures* (for Toronto and other Canadian stations). Newspaper reports, on the other hand, present such extreme temperatures as a discrete set of information on successive heat wave events. This helped us in studying heat waves in Delhi and Toronto during the past several decades. **Second**, hazards of heat waves, like those of cyclones and floods, constitute a major human interest topic for the news media. The journalists who are masters of communication excell in symbolic production of such human interest topics. One of our objectives is to interpret how newspaper reports have employed different types of symbols and metaphors to describe human sufferings resulting from heat waves in two different cultural contexts. **Third**, climate change has emerged as one of the central issues in global media discourse on recent heat waves throughout different parts of the world. In such discourse there are frequent references to the scientific literature, including those cited in IPCC reports, attributing heat waves and other climatic anomalies to global warming and climate change. One of our objectives is to assess how climate change has been implicated in some of the newspaper reports on recent and past heat waves in Delhi and Toronto.

Climatic similarities and differences between Delhi and Toronto

Comparing heat waves in Delhi, India with that in Toronto, Canada may seem like a comparison between apples and oranges. Yes, the differences are striking but there is at least one similarity, as we describe it below.

Both have a continental climate

Both Delhi and Toronto are situated in the interior of the continent. Delhi is situated nearly 1000 km from the coast of the Indian Ocean. Toronto is situated several thousand kilometers from the Pacific Ocean and several hundred kilometers from the Atlantic Ocean. Climates of such interior places are called *continental climates*, which are characterized by large differences in temperatures between the summer and the winter months. Such differences in seasonal temperatures are often called *annual ranges of temperatures*. Irrespective of their differences in latitudes, continental climates tend to have hot or very hot summer temperatures. As Delhi is situated in a subtropical climate at 28.6° N latitude (and 77.2° E longitude), it has a much warmer climate than Toronto. Yet, it experiences considerable ranges of monthly temperatures within a hot climate, owing largely to its interior location. Although Toronto has a much colder mid-latitude climate, as it is situated at 43.7° N latitude (and 79.4° W longitude), that is, several latitudes and more than 1000 km north of Delhi, it still experiences a relatively hot summer, largely owing to its interior location and the resulting continental climate. Toronto has a true continental climate with an extremely cold winter and a warm to hot summer. Some of the details of temperature and precipitation characteristics of both of these cities are summarised below in Tables 7.1 and 7.2.

Climate of Delhi

According to Köppen's Climatic Classification, Delhi has a *humid subtropical climate* (Köppen's symbol: Cwa). This is a formal textbook climatic type based on certain specific criteria for humid climates. In reality, the term "humid" is misleading because Delhi has essentially an arid climate for two-thirds of the year (October to May) when each of these eight months receives very light to nearly trace amounts of rainfall, ranging between 6 mm (0.25 inches) and 30 mm (1.2 inches). In all, this prolonged dry season receives only 14% of the total annual rainfall of 795 mm (31 inches). The rainy season begins in the month of June with light rainfall (about 54 mm or 2 inches). At least 75% of the annual rainfall (about 600 mm or 24 inches) occurs during the months of July, August and September. Thus, the rainfall regimes of Delhi are extreme with a heavy concentration of summer monsoon rainfall which is followed by a prolonged dry period.

Annual temperature regimes of Delhi are equally extreme. It is controlled by its low (subtropical) latitudes and continental (interior) location (as introduced briefly above). Because of these twin effects Delhi has a hot climate, with an average monthly maximum temperature of 32°C (89°F) and monthly minimum temperature of 19°C (66°F). The average January temperature for a period of 100 years was 7.8°C (46°F), but in some years coldest daily temperatures in December and January had dipped down to 5°C (41°F) or lower due to the effect of *advection* (transport from one area to another) of cold winds from the Himalayas. In extreme cases, below-freezing temperatures (-1°C to -2°C) had also been recorded. In contrast, high temperatures throughout the warmest six months (April-September) dominated the stories of Delhi's heat waves. The *average daily maximum temperatures* during 1956-2000 ranged from about 37°C (99°F) in April to 40°C or more (nearly 105°F) in May and June and to 34°C (93°F) in September. Even the *average daily minimum temperatures* (i.e., the night-time lows) during this period exceeded 25°C (76°F). As comparative data in the first and the second columns of Table 7.1 indicate, average daily maximum temperatures for the entire year, especially for the summer months, have increased during the last half of the record (1956-2000). Most of the heat waves occurred during the month of May when the daily maximum temperatures often reached extreme values such as 45°C (113°F) to 46 °C (115°F).

Climate of Toronto

According to Köppen's Classification, Toronto has a *humid continental climate* with a warm or hot summer (Köppen's symbols: Dfa and Dfb). In this case, the term *humid* is more applicable for Toronto than for Delhi because its annual total precipitation of 786 mm (31 inches, i.e., almost identical with Delhi's total rainfall) was distributed fairly uniformly throughout each month, no month receiving less than 48 mm (1.88 inches). There was a summer concentration of rainfall in five months (May to September) but, unlike Delhi's heavy summer concentration (about 75% in three months), only about half (48%) of Toronto's total precipitation occurred during this period. Each of these five months received between a minimum of 72 mm (2.8 inches) and a maximum of 78 mm (3 inches). One of the obvious characteristics of Toronto's precipitation that differed from Delhi's rainfall regime was that nearly one-third of its annual precipitation occurred as snowfall during six months (October to March). For this reason, this type of mid-latitude climate is often characterized as a snow climate.

Annual temperature regimes of Toronto provide even greater contrasts with that of Delhi. Its average daily maximum temperature (based on a 30-year period) exceeded 20°C (68°F) in only four summer months (June, July, August, September). The average daily maximum temperature in the month of July was 27°C (80°F). The daily minimum temperatures in summer months ranged between 7°C (about 45°F) in May and 16°C (64°F) in July. Winters were bitterly cold as the average daily minimum temperature dipped down to -9°C (15°F) in January. On the contrary, the *extreme daily maximum temperatures* (not reported in the table) ranged between 31.1°C (88°F) in April and 38.3°C (101°F) in August. The extreme daily maximum temperature of 38.3°C (101°F) was recorded on 25 August 1948. Heat waves in Toronto could thus occur in anyone of the summer months.

What is a heat wave?

The Indian Meteorology Department (IMD) and Environment Canada (EC) define heat waves significantly differently. From these definitions it becomes clear that the residents of Delhi are used to much hotter air temperatures of its tropical/subtropical climate than those living in a colder mid-latitude climate of Toronto.

A *heat wave* in India is defined by the IMD as a period (unspecified) when the maximum temperature of a station reaches 40°C (104°F) for

plains and at least 30°C (86°F) for hilly regions (http://www.imd.gov.
in/ . . . retrieved on 4 May 2015). Although the IMD acknowledges
that a combination of high air temperature and relative humidity may
result in uncomfortably "hot days", it does not include humidity in
its determination of heat waves. Instead, the emphasis is on deviation
of air temperatures from *normal* (long-term average) when the station
temperature reaches 40°C (104°F). Thus, a *severe heat wave* is a departure
(increase) of 6°C to 7°C above a "normal" daily temperature of 40°C
(104°F).

The meteorological branch of Environment Canada defines a *heat
wave* as a period of three or more consecutive days when the maximum
temperature reaches 32°C (90°F) or more (https://ec.gc.ca/meteoaloeil-
skywatchers/default . . . retrieved on 4 June 2015). In addition to air
temperature, Environment Canada combines relative humidity with
air temperature to calculate a unique index called *humidex*, to assess
human discomfort (how people feel) due to hot weather. Because
of the compounding effect of humidity with high air temperature
the humidex values (without any unit) are always higher than the
original air temperature (7.3). A *heat warning* (heat alert) is issued by
Environment Canada using a one-day-forecast of humidex over 40.
Human discomfort is further classified by EC based on the following
ranges of humidex values:

- Humidex: 30-39: some discomfort
- 40-45: Great discomfort (avoid exertion)
- >45: Dangerous discomfort

2015 May heat waves in India

There were 223 heat waves in India between 1978 and 1999,
resulting in at least 5300 deaths (Murari and others 2015). India's
worst heat wave occurred in 1998 when 2541 people lost their lives
due to heat-related complications, such as sunstroke, dehydration, and
gastrointestinal disorders (http://www.indiaenvironmentportal.org.in/
media/iep/infographic/ . . . retrieved on 25 June 2015). The second
worst heat wave occurred in 2015 with a fatality of at least 2000. Some
of the newspaper reports have reported a higher figure, such as 2360

by *China Daily* (9 June 2015; see Table 7.6). The Indian Meteorology Department (IMD) has further provided data indicating that the past decade (2001-2010) was the warmest decade on record with a temperature anomaly of 0.49°C. Both Indian climate scientists and the IPCC warn of increasing heat waves in future in India and many parts of Asia. The month of May, 2015 proved to be such a historic period for heat waves throughout India as at least ten of India's 29 states (eleven including Delhi Union Territory) experienced severe heat waves (Table 7.4). At least 50%-90% of the areas of these states experienced heat waves ranging between 40°C (104°F) and 45°C (113°F). In the remaining 10%-50% of the areas of these states heat waves exceeded 45°C (113°F). It can be seen from data in Table 7.4 that the states which were hit hardest included Rajasthan, Punjab, Haryana, and Uttar Pradesh.

The 2015 heat wave was particularly problematic for Delhi, which experienced some of the longest spells of heat waves, as data in Table 7.5 indicate. There were two successive heat waves: the first wave lasted for six days (5-11 May), whereas the second one prolonged for two weeks (18-31 May) (Table 7.5). The first wave was marginally less severe than the second wave. The threshold value of 40°C was recorded on 5 May. During the second wave the daily maximum temperature exceeded 45°C on several days. The heat wave was compounded by the complete lack of rainfall and low atmospheric humidity during these episodes (Table 7.5). Humidity during early morning hours hovered above 50% but by early afternoon the heated air became extremely dry when relative humidity fell below the tolerable limit of 50% to as low as 18-22% (Table 7.5).

Discourse analysis of newspaper reports on heat waves in Delhi

Digital copies of newspaper reports

Digital copies of newspaper reports provide a rich source of historical information on heat waves in India, including numbers of heat-related deaths and geographical areas most severely affected by heat stresses. We used *LexisNexis® Academic*, an electronic database (a software/program for searching news documents) as the primary tool for retrieving past copies of newspaper reports on heat waves in Delhi. Our initial search using the phrase (search words) "Heat waves in Delhi, India" produced the following results:

- Newspaper reports: 877 (or higher depending on a specific date)
- Web-based publication: 105
- Newswires and Press Releases: 14
- Industry Trade Press: 10
- Magazines and Journals: 4
- Others: the expanded list is not reported here

We were interested only in newspaper reports, specifically in discourse analysis of these reports.

What is discourse analysis?

The simplest meaning of discourse is the language of communication or expression. In this study we define discourse analysis of newspaper reports as a *systematic analysis of texts of news reports dealing with different subject matters, especially typified by recurrent themes.* In a broad sense, discourse analysis is a form of content analysis but our methodology differs significantly from most of the conventional techniques of content analysis. Whereas the latter may use a numerical (quantitative) description of certain structures of the text (such as key words and sentence structures), often using computer-assisted techniques (including different types of software), we used a descriptive method for identifying a number of recurrent themes (media messages) called *media frames* or *news angles*, or simply *story lines* (Hannigan 1995 and 2014, Rashid 2011). The news media tend to employ such story lines on a routine basis to organize different aspects of a news event. The story lines help both the reporter (journalist) and the general reader (public) in sorting out major relevant themes out of chaotic news.

Judgement sample

We have identified a selected number of media frames by screening all of the 877 reports in two successive steps. At the first stage we browsed through the reports in search of recurring themes (frames). Next, we selected a "judgment sample" of 25 reports (Table 7.6) for discourse analysis, based on the following criteria:

- Relevance (heat wave-related news)
- Reports with data on air temperatures and fatality
- Reports with heat wave impacts and coping strategies
- Reports with other human interest topics

We are characterizing this small sample as a "judgment sample" because we used our judgment in selecting them based on the above criteria without employing a more rigorous statistical method of sampling. Yet, our search was not entirely arbitrary as we were guided by two basic principles. **First**, we wanted to cover as many years as reported between 1972 and 2015. Based on the IMD data (at least 223 heat waves during this period), it seems that the *LexisNexis* database probably included reports only on some of the major heat waves. This explains why some of the years are missing in Table 7.6. **Second**, we have chosen only those reports which contained the most complete set of information on air temperature and fatality (latest updated numbers). Thus, for example, if there were three or four successive reports within a range of few days or on the same day or on the same event we had chosen the report that had the latest and the most complete set of information on air temperatures and fatality.

Routine information on air temperature and fatality

One of the special characteristics of newspaper reports on heat waves was that they provided information in a format that was not readily available elsewhere, including government websites. The majority of reports contained two types of routine information on heat waves: (a) air temperature and (b) number of heat-related deaths. Of these, data on air temperatures for a given date were fairly consistent among different newspapers because these were based on the IMD source. In this regard, the 1972 *New York Times* report on the *daily maximum temperature* at Delhi (100°F or 37.8°C) seems to be an exception. The remaining data indicate that the daily maximum temperatures at Delhi for each of its heat wave events exceeded the threshold value of 40°C (104°F), mostly

within the range of 40° to 45°C (113°F). The 2015 May heat wave seems
to be a record for Delhi, as the Indian newspaper *Siasat Daily* reported
that a Delhi experimental station registered the maximum temperature
of 50°C (122°F). The official temperature data for Delhi normally refers
to its weather station at Palam.

Another finding on record air temperatures in India is significant.
It may be stressed here that although Delhi (including New Delhi, the
national capital of India) is one of the hottest places in India, almost
invariably several other states were even hotter than Delhi. As data in
the third column of Table 7.6 indicate, the daily maximum temperature
records at some of the other states ranged between 45°C (113°F) and
a whopping 51°C (124°F) (Note the exception of 2014). Among the
places that experienced such extreme maximum temperatures, the
leading states were Rajasthan, Punjab, Haryana and Uttar Pradesh.

Most of the heat-related deaths were due to stroke and dehydration.
The numbers of deaths reported in different newspapers often varied
from one another, partly based on different dates of reports and partly
based on data sources. Overall, data on heat-related deaths should be
treated as tentative because even government sources kept on revising
their data, depending on the latest updates and estimates. To explore if
our samples of newspaper reports (Table 7.6) represented a significant
proportion of data on heat wave deaths in India between 1978 and 1999
(dates for IMD estimates), we have added all data (latest data) in Table
7.6 for the same period (excluding repeat data for the same year). The
total numbers of deaths (4366), reported in the newspapers, represented
82% of the total deaths (5300) estimated by the IMD sources (Murari
and others 2015). Another verifiable finding from our sample is that,
with a fatality figure of 2360 (official estimate: more than 2000), the
2015 May heat wave was the second mostly deadly heat wave in India
in the last 40 years, following the 1998 heat wave which was responsible
for 2541 deaths.

Geography of heat waves based on The Times of India *reports*

The *LexisNexis* database indicated that out of 877 newspaper reports,
The Times of India (TOI) published 115 reports between 2010 and 2015.
When we browsed through this list we found that some of these reports

were by other newspapers (i.e., reproduced in TOI). For our purpose, we found that a sample of 25 TOI reports provided an adequate source of information for the geography of heat waves in India (Table 7.7). Data in Table 7.7 indicate that heat waves have become a pervasive national climatic hazard in India, as 17 of India's 29 states experienced major heat waves during the last five years (2010-2015). Based on frequencies of reports, four states experienced most severe heat waves: Rajasthan, Uttar Pradesh, Haryana and Punjab. This five-year trend is consistent with the single-day heat record on 25 May 2015 (Table 7.4), which indicates that the same four states experienced the most severe heat waves. The NOAA map from which we have interpreted all data in Table 7.4 does not show Delhi, which is also a union territory (like Canberra, Australian Capital Territory, short-formed as ACT). Delhi Union Territory has a relatively small area compared to other states. Probably for this reason it was not shown on the NOAA map (hence not in Table 7.4). One of the striking characteristics of recent expansion of heat waves in India is that some of the unlikely places, such as the high-altitude hilly states of Himachal and Uttarkhand, have also been experiencing heat waves (Table 7.7).

The maximum daily temperatures during the last five years, reported in TOI (Table 7.8), seem to indicate that most of the records of extreme temperatures are continuation of the past trend for the last forty years. Thus, the record temperatures for Delhi and other Indian states are similar for both of the periods, i.e., 40-year trend (Table 7.6) and last 5-year trend (Table 7.8). The latest data confirms that Delhi is one of the hottest places in India, as its maximum daily temperatures during the last five years (2010-2015) ranged between 42.2°C (108°F) and 47.6°C (117.68°F) (Table 7.8), not much different from the range of 44°C (112°F) to 46.1°C (115°F) during the longer period (1978-2009) (Table 7.6). The five-year data also confirms, similar to the longer-term data, that some of the other places in India were even hotter than Delhi. Most often these places were in one of the four hottest states, namely Rajasthan, Punjab, Haryana and Uttar Pradesh. The record maximum temperatures at some of these states during the last five years reached 48°C (118.4°F) or even slightly higher (Table 7.8). The record temperature of 50°C (122°F), reported by the *Siasat Daily*, India (Table 7.6), is an exception for an experimental station and it is not the IMD official record.

Problems of daily life due to heat waves

Besides routine information on air temperature and heat-related fatality, some of the media frames dealt with impacts of heat waves and different types of coping strategies. The following are examples of adverse impacts of heat waves on daily life:

Problems for the poor (*The Washington Post*, 14 June 1994): "It's been a very bad summer", said a 25 year old construction worker who was carrying bricks on his head in today's 98-degree heat. "It is especially difficult for laborers. Sometimes everything goes hazy before my eyes, I get dizzy. I constantly have to keep watch on myself".

Dehydration and water-borne diseases (*The Times of India*, 19 May 2013): Medical experts say that incidences of senior citizens suffering from dehydration have also increased. "We are getting several cases of food and water borne diseases. People should not to eat outside or refrigerate food for long at home", said a senior doctor.

Sweats and heatstroke (*The Times of India*, 11 May 2011): All over the city, the impact of the heat wave could be clearly seen as Delhiites sweated in the sun and complained of heatstroke. "I had to walk from outer circle in Connaught Place to inner circle where my car was parked in the afternoon and by the time I got there, I was drenched in sweat. Even switching on the air-conditioner on full did not help much because it was just so hot", said a businesswoman.

Power overloading and blackout (*The Scotsman*, 12 June 1995): "But the majority are resigned to suffering in silence as power stations burn out due to overloading. Others illegally tap into power lines, overloading the system and causing burnouts which often take days to repair. Several areas across Delhi are without power, often for 12 hours at a stretch, forcing people to run generators which spew forth diesel fumes, adding to the overall pollution and heat".

Water shortages (*The Scotsman*, 12 June 1995): "Water taps run dry even in the city's pampered, tree-lined suburbs inhabited by politicians, senior officials and other VIPs. Water supplies are down to a trickle, despite the promise of additional resources from rivers in the neighbouring states of Haryana and Uttar Pradesh following appeals by the prime minister, Narasimha Rao. Huge tankers are doing brisk business selling buckets of water in less privileged areas as hand pumps and pumping stations run dry".

Delhi streets as "mini battlefield" (*The Scotsman*, 12 June 1995): "Driving down Delhi's potholed roads, fanned by hot, dust-laden winds from the north-western desert areas, is torture. Snarl-ups in the city's traffic—which even on normal days is unbelievably chaotic—have become mini battlefields, where abuses and fists fly at the slightest infringement".

Coping strategies

Stories on coping strategies varied widely from one report to another.

Fewer people in offices (*The Times of India*, 2 July 2012): Every day, fewer and fewer government employees appear at their offices.

Schools closed (*The Times of India*, 2 July 2012): All schools, including government, private-run and those under civic bodies, which were to reopen on Monday after the summer vacation, have been told to extend the vacation till July 8. Now the schools will reopen on July 9".

Stay indoors (*The Times of India*, 19 May 2013): "I had plans to go out for lunch with friends but the weather was so bad that we decided to stay indoors. Power cuts for even a minute in this weather are unbearable", said a resident of Delhi.

(*The Times of India*, 18 May 2010): ". . . it was impossible to spend more than a few minutes out in the open. The sun was extremely harsh and I had to cancel all my afternoon appointments so I could stay inside my air-conditioned office", said a resident of Patparganj.

Stay cool by immersing in river water (*The Scotsman* 12 June 1995): "Some Delhi residents have taken to spending their days immersed waist-deep in the Yamuna River which runs through the city. They prefer to share the polluted, muddy water with buffaloes, cows and, at one spot, even elephants rather than brave the searing heat elsewhere. Even society women have taken to showering in their saris, hoping the layers of the wet cloth will give them some temporary relief from the suffocating heat".

Selling glasses of cool water (*The Washington Post*, 14 June 1994): For others, the heat has been good for business. Hari, who uses only one name and sells glasses of cool water for 25 paises (less than a penny) near a busy city marketplace, noted "At first I was sad about all the people dying in the heat. But when there is such a lot of heat, my business goes up".

Power saving devices (*The Times of India*, 28 May 2015): With meteoric increases in demand for power many power companies provided power-saving incentives to consumers. One such incentive was that each registered consumer could get up to four LEDs at subsidized rates, thereby saving 125 kWH per year. Nearly half a million customers took up the offer. It was a win-win scheme for consumers who would be able to save up to Rs 200-250 (about $5 per bill) in power bills by switching to LEDs.

Cooling centres (*The Times of India*, 1 May 2024): Public Health Foundation of India is developing 'Heat Action Plans' for cities, starting with Delhi. Highlights of the plan include setting up cooling centres in mosques, temples and other public spaces; training employees at local hospitals and urban health centres to deal with heat-related emergencies, and teaching people how to stay cool. PHFI had developed a heat action plan for Ahmedabad being implemented by the municipal corporation there since last year.

Climate change implicated

The latest climate change models project that, by 2070-2099, the intensities, frequencies and durations of heat waves in India are expected to increase significantly, especially in northern India (Murari and others 2015). Many Indian scientists and government officials are familiar with these projections and, therefore, attribute current heat waves in India to this ongoing process of global warming. Echoing this story line, some of the newspaper reports attributed heat waves in India to climate change.

Definition of heat waves: Higher threshold temperatures (*The Times of India*, 14 May 2010): "The definition of heat wave has changed from last year", informed an RMC official. "Earlier if the temperature went five degrees above normal or average temperature for over three consecutive days, we used to term it a heat wave. Now, if the temperature crosses 45 degrees, it's a heat wave".

Increases in Delhi average temperatures (*The Times of India*, 30 May 2012): The mercury has touched (at Delhi) 45 degree C mark five times during the month. In the last five years, it happened once in 2011, twice in 2010 and the mercury struggled to touch the figure in 2009, 2008 and 2007. It has also lifted the mean (average) temperature of May, which used to hover around 40 degree C mark. This year, the figure is close to 43 degree C, at least three degree higher than the average temperature of the month.

Causes of heat waves (*The Times of India*, 21 May 2013): "The heat wave is the result of the absence of any western disturbance activity in Delhi and neighbouring areas, strengthening of hot northwesterly winds from the desert and subsidence of air in association with an anticyclone over Rajasthan and adjoining areas," said the deputy director of meteorology, Delhi regional meteorological centre.

IPCC report on heat waves in Asia (*The Times of India*, 1 May 2014): . . . "the Intergovernmental Panel on Climate Change's recent report that highlighted how "frequent and intense heat waves in Asia will increase mortality and morbidity".

Global warming evident (*The Times of India*, 29 May 2015): Climate records show that human-induced global warming had turned 2014 into the hottest year on record. Eight out of the 10 warmest years in India were during the recent past decade (2001-2010), making it the warmest decade on record with a decadal mean temperature anomaly of 0.49 C. "The number of heat wave days may go up from about 5 to between 30 and 40 every year", added CSE climate researchers.

Urban heat island (*The Times of India*, 29 May 2015): Cities feel the brunt of the elevated temperatures, because of the magnified effect of paved surfaces and a lack of tree cover - this is known as the "urban heat island effect".

Symbolic news

Journalists excel in presenting human interest topics by using metaphors and symbols (Rashid 2011). News reports were replete with such symbolic news. Here are some examples:

"Killer heat" (*Times of India*, 26 May 2010): "It's the peak of summer, a time when heat waves are common in most parts of India. But what the country is currently witnessing is killer heat that has shattered temperature records in many cities and has in its grip an entire swathe from . . ."

"Roast pit" (*Times of India*, 27 May 2015): "The capital remained on the roast pit for the fifth straight day, with the maximum temperature a fiery 45 degrees Celsius on Tuesday".

"Testing India's mettle" (*The Scotsman* 12 June 1995): **"47C heat wave tests India's mettle"**.

"Hotter than Saudi Arabia's Empty Quarter" (*The Independent*, London, 4 June 1994): "It is now hotter on the streets of Jaipur or at the Taj Mahal in Agra than it is among the dunes of Saudi Arabia's Empty Quarter (called *Rub-al Khali* in Arabic, for the largest desert in eastern Saudi Arabia).

"Dust but not rain" (*The Independent*, London, 4 June 1994): ". . . and many Delhiites seem to spend an inordinate time on their roofs, scanning the sky for an approaching cloud. But lately, the sky is full of dust, not rain".

Heat waves in Toronto

Canada has a cold and snow climate. Talking about heat waves in such a climate may seem like an oxymoron. Yet, summer heat waves are common in several places throughout Canada. There are many climatic processes which are responsible for such heat waves but most of the places that experience heat waves have continental climates, that is, they are situated in the interior of the continent. For example, Saskatoon (Saskatchewan) and Winnipeg (Manitoba), situated in the interior Prairie Provinces of Canada, experience some of the coldest winter temperatures in the world. Yet, they have the surprising distinction of experiencing some of the highest numbers of *2-day heat waves* in Canada during a period of nearly one century (Smoyer-Tomic and others 2003). Toronto has a much milder climate than that of Saskatoon or Winnipeg, but it has the distinction of having the record of experiencing the highest numbers of heat waves per year: a total of 249 days during a period of six decades (1938-1998), consisting of combined values of 2 days, 3 days, 4 days, and 5+ days of heat waves (Table 7.9). (See Smoyer-Tomic and others 2003 for details on heat waves in 25 Canadian cities).

It should be stressed here that Environment Canada (EC) recommends 32°C as the absolute threshold for *heat alert*, i.e., without including the relative humidity values. This may be justified based on the historical record which indicates that during many past heat waves the *maximum daily temperature* at Toronto reached 32°C or higher for several days in a row. For example, the record highest maximum daily temperature at Toronto—38.3°C (101°F) on 25 August 1948—was a part of a six-day heat wave exceeding 32°C (Table 7.10).

On further analysis, we find that EC's threshold (32°C) is too restrictive in the sense that it misses out wide ranges of data on humidex with "some discomfort". Smoyer-Tomic and others (2003), on the other hand, have used a threshold of 30°C, which seems to be consistent with a frequently used definition of heatwaves in public/newspaper discourse. This definition makes more sense because, based on the daily maximum temperature of 30°C, humidex values may reach 34 ("some discomfort") with a relative humidity (RH) of 40%, or 40 ("great discomfort") with an RH of 65%, or as high as 46 ("dangerous" humidex) with an RH of 90%. Hourly humidex values on 17 July 2012 provide an example (Table 7.11).

Another recent month with record high maximum daily temperatures was July 2011. The daily maximum temperatures in this month reached or exceeded 30°C (86°F) on 15 out of 31 days. There were several heat waves (Table 7.12). Hourly humidex values (published by Environment Canada dating back to 1953) indicate that during this period of record (1953-2015) there were many heat waves with high humidex values. The past decade, in particular, was one of the warmest decades with several continuous hours of high humidex values. One such record belongs to 21 July 2011(Table 7.13) when humidex values for Toronto ranged between 42 and 48 ("great discomfort") for 13 hours (8 a.m. to 9 pm) and "dangerous" humidex values (46-48) for a continuous period of six hours (11 a.m. to 5 p.m.).

Based on our review of samples of similar data for the past six decades (not reported here), it is clear that Toronto has been experiencing many episodes of heat waves at threshold temperatures of 30°C (or slightly higher), that is at about 10°C lower threshold than that of Delhi, India.

Discourse analysis of newspaper reports on heat waves in Toronto

Sampling methods

As in the case of Delhi, we used *LexisNexis® Academic* program as the primary tool for retrieving past digital copies of newspaper reports on heat waves in Toronto. Our initial search, using the phrase "Heat waves in Toronto, Canada" produced a list of 991digital copies of newspaper reports published between 1981 and 2015. By far, the largest numbers of reports were published by *The Toronto Star* (635 reports), followed by *The Globe and Mail* (96 reports) and *National Post* (49 reports). For discourse analysis our goal was to obtain a sample of 25 reports, comparable to the Delhi sample. However, we adopted a more thorough screening method for obtaining this sample in two successive steps. **First**, we browsed through all of the 991 reports and obtained an initial "judgment sample" of 121 reports based on the identical criteria described for the Delhi sample (see earlier text). **Second**, for our detailed discourse analysis of newspaper reports we then used a *random sampling* method for obtaining a blind sample of 25 reports out of the initial sample of 121 reports. For this, we used a random number table between 1 and 121, which were obtained from a web-based random number generator (table) (*http://stattrek/statistics/random-number-generator.aspx* . . . retrieved on 7 July 2015).

Nature of discourse analysis

Following a review of all of the 25 reports we identified six major recurring themes (media frames) as follows (Table 6.14):

HWD: Heat wave details
HLI: Health-related issues
POP: Power problems
OTH: Other problems
COP: Coping strategies
CCI: Climate change implicated.

Following the selection of the above themes, we conducted a comprehensive discourse analysis of our random sample of 25 reports, by classifying each of the reports into a number of segments, each containing one or more of the above concepts. Each segment consisted of either one paragraph or several paragraphs or a paragraph and a separate sentence (or several paragraphs and several sentences). By using the Microsoft word-count function, we then counted the numbers of words in each classified segment. The numbers of words were entered into an Excel spreadsheet for subsequent tallying. The lengths of the reports varied between 127 words and 1078 words, with an average of 482 words. The relative proportions of each of the concepts in a given report were calculated as percentages of the total numbers of words in a report (Table 7.14). In all, we classified about 12,401 words in 25 reports. The summary data (last row in Table 7.14) indicate the proportions of space (percent of total words) devoted to each of the six recurring themes.

Routine information on maximum daily temperatures

Although maximum daily temperatures were the building blocks of all news on heat waves, we found that much fewer reports on heat waves in Toronto contained actual data on air temperatures compared to the equivalent sample of 25 reports on heat waves in Delhi (Table 7.6). Compensating this limitation, Toronto data had one major advantage in that these could readily be compared with the official data published by Environment Canada (on its websites) (Table 7.15). We could not access similar data for Delhi (i.e., data on daily temperatures) through the IMD websites.

One of the reasons for a limited number of useable data ("good data") on heat waves in Toronto (reported in only nine out of 25 reports) was that some of the newspaper data were not comparable to EC data because of many different forms of journalistic expressions of temperature data. Among a variety of expressions, the following were common:

- Often newspapers reported a record temperature on a given date which might not have coincided with the date of publication. Sometimes the references were not clear and we had to interpret the actual date of record.
- Sometimes newspapers reported on the maximum temperature on a past date, such as "in this weekend" for a report published on Monday or Tuesday (for example).
- Some of the reports published expected maximum temperature for a future date (for example for tomorrow). We did not include such data.

As comparative data in Table 7.15 indicate, most of the useable newspaper-based data matched closely with official data. Only in some cases, there were minor deviations. We are not sure if these were related to publication errors or to the preliminary nature of the real-time data (i.e., the reported data) from government sources. It is possible that some of the preliminary data were revised subsequently by Environment Canada before publishing them on their websites.

Details of heat waves (HWD)

As data in the second column (under HWD) of Table 7.14 indicate, about three-quarters of the reports (18 out of 25) devoted some space (amounting to about 20% of the total words) for describing the nature of heat waves. The *Toronto Star* (26 June 1989) and the *National Post* (2 August 2006) devoted their entire space (100%) to the details of heat waves. Here are some excerpts from the *Toronto Star* report:

> Today's' high temperature forecast was for 30 Celsius (86 F) for the third day in a row, and tomorrow should be another hot one . . . Though this weekend wasn't as sizzling as the average 35C (95F) temperatures that marked last year's heat waves, high levels of humidity— around 60 to 85 per cent—made it feel like it was 36C.

> The hottest June 26 in downtown Toronto was 34.4C (94F) back in 1952.

> While Toronto was the hottest major city in Canada this weekend, the tiny town of Happy Valley-Goose Bay, Labrador, beat us by 5 degrees with an average weekend temperature of 35C.

Health issues (HLI)

Unlike newspaper reports of large numbers of heat wave-related deaths in Delhi (about 5300 deaths between 1978 and 1999), very few such deaths in Toronto were reported by newspapers. Although we expected this because of the much lower intensities of heat waves in Toronto than in Delhi, yet newspaper-based data did not match with official sources (reported in Pengelly and others' 2007). The latter suggested that Toronto's heat-related deaths between 1938 and 1998 were about 120 per year. Perhaps this discrepancy resulted from fundamental differences in methodology for estimating heat-related fatality. Whereas official sources (reported by Pengelly and others) included pre-existing medical conditions which were exacerbated by abnormally high air temperatures, these were difficult reporting problems for the news media. Some of the difficulties of reporting mortality data were stated directly in a *Toronto Star* report (20 July 2005): Six elderly deaths were investigated if these were related to heat waves because "often it isn't solely the heat that kills but a pre-existing illness aggravated by rising mercury levels . . . The attending physicians needed to find out about their medical history". Perhaps, because of such uncertain data on mortality we found very few references to actual heat-related deaths, as the following examples show:

- 1988 June 25 (*The Toronto Star*): 1 death
- 1988 July 13 (*The Toronto Star*): 19 deaths investigated
- 2001 August 1 (*The Toronto Star*): 2 senior deaths due to heat stroke
- 2001 August 11 (*The Globe and Mail*): 4 deaths
- 2005 July 13 (*The Globe and Mail*): 4 deaths
- 2005 July 20 (*The Toronto Star*): 6 deaths investigated

It was much easier for journalists to report on different types of health issues (HLI). The bulk of the 13% text on this item (column 3 in Table 7.14) dealt with two main problems: (a) different phases of heat exhaustion and (b) health issues for the elderly. The earliest report on heat exhaustion was published in the *Globe and Mail* on 10 July 1981 in its news "in brief" (127 words). An excerpt from this report:

> [According to a medical expert] Heat cramps, which can be cramps of any muscle, can affect even young, healthy people, particularly if engaged in vigorous activities. The major symptom of heat exhaustion is a feeling of weakness, often accompanied by headache or nausea. Rest and fluids usually cure heat cramps and heat

exhaustion. Heat stroke, which is rare, is an emergency. The body loses its ability to control temperature, blood pressure can drop dangerously and the heart can stop.

Out of the eleven reports on health issues four reports dealt specifically with the seniors. For example, the *Toronto Star* report on 13 July 1988 devoted its entire text (100%) to heat-related deaths of 19 seniors with the following banner: "Ontario coroner reviewing seniors' deaths in heat waves". It further reported:

- Several seniors were treated for heat shock at Metro hospitals during the heat wave when temperatures exceeded 37C (98F) for several days. Many of the patients had dangerously high core body temperatures, some reaching as high as 42.1C (107.8F). Normal body temperature is 37C (98.6F) . . . More important than arriving at an exact number of heat-related deaths will be analysis of why some died, and the drafting of measures that could prevent future fatalities . . . [said a medical expert].

Power problems (POP)

About 10% of the total text (covered in nine reports) provided some of the details of power problems resulting from excessive demand for electricity, largely due to heavy consumption of electricity by air-conditioners and fans. A *Toronto Star* (15 July 1987) report devoted its entire space (100%) to a power blackout in Toronto with the following heading: "Power failure strands tenants nearly 24 hours". Here is a small segment of that report:

> "Tenants at twin midtown high-rise apartments that were paralyzed for almost 24 hours by a power failure are upset that no emergency lighting system was available to help them find their way out.
>
> Power was finally restored about 10.30 last night for the 800 residents of 30 and 50 Hillsboro Ave., who were faced with pitch-black corridors and stairwells. Water

was cut off to those living above the 10th floor in both
buildings".

Excessive demands for electricity had always been a problem for
Ontario Hydro (company responsible for electricity supply) during
successive years of heat waves. The *Toronto Star* reported such problems
during previous episodes of heat waves, such as in 2004 and 2006.
More recently, the company seemed to be in a better position to handle
demands for power, as the *Star* headline on 21 June 2012 implied:
"Toronto heatwave no sweat for hydro; electricity supply is up while
demand is down this balmy town around".

Other problems (OTP)

There were only four reports dealing with some of the secondary
issues related to heat waves, but two of them provided details of the
following problems:

- *The Globe and Mail* (9 July 1988): (a) Industrial shutdown due
  to high levels of air pollution, and (b) forest fires in northern
  Ontario.
- *The Toronto Star* (12 July 1988): Water supply problems in
  Toronto due to excessive draw-down of its reservoirs.

Coping strategies (COP)

By far the largest amount of space was devoted to coping strategies
(26% of the text). As a relief from the heat-wave, people crowded
beaches and swimming pools. Some spent time in air-conditioned
movie theatres. There were predictable increases in the sale of some of
the cooling devices, notably electric fans, air-conditioners, sprinklers
and wading pools. Sale of beer and ice cream also increased with
successive heat wave events. Among all coping strategies, perhaps, the
most popular measure included taking shelter in different types of
air-conditioned buildings. Several reports of the *Toronto Star* listed the
following "cooling centres":

- Ontario Place (18 June 1994)
- (Unspecified) Air-conditioned buildings (20 June 1995)
- (Idea of) Metro Hall as a cooling centre (14 July 1999)
- Shopping malls and libraries (2 July 2002)
- Formal cooling centres throughout Toronto (25 June 2003)

The idea of a formal cooling centre in the Metro Hall was first proposed by the late Jack Layton (the former NDP leader), who was a Councillor of the City of Toronto in 1999. He wanted Lastman (Toronto's Mayor at that time) to "start a heat watch and air pollution warning system, and operate it from his office like Chicago Mayor Richard Daley".

Chicago's cooling centres were introduced after the 1995 heat wave. Public buildings in every neighbourhood with air conditioning had been designated as places where residents could stay during a heat emergency (*The Toronto Star*, 14 July 1999). The mayor was not enthusiastic as he cautioned: "Metro Hall is not set for this stuff. Metro Hall does not have the toilets, does not have the facilities, does not have anything". He said he'd consider taking some kind of action if the temperature climbed over 32°C but only as a temporary measure and never on a daily basis. Despite this initial setback, the proposal for a formal cooling centre in the Metro Hall was finally approved by 2003 as it became one of the four cooling centres of Toronto:

> *The Toronto Star* (25 June 2003): The heat alerts led to the opening of the four cooling centres. More than 80 people dropped into the centre at Metro Hall, at 55 John St., yesterday. The others are at the Etobicoke Civic Centre, 399 The West Mall; East York Civic Centre, 850 Coxwell Ave.; and North York Civic Centre, 5100 Yonge St.

Climate Change Implicated (CCI)

As in the case of Delhi, most of the climate models project significant intensification of heat waves in Toronto by the end of this century. According to climate models, based on the "IPCC middle-of-the-road emissions scenario", Toronto is expected to experience above

30°C temperatures for about 30 days per year by 2020-2040 and for at least 50 days per year by 2080-2100, according to a Government of Canada (Health Canada) study. These are relatively recent projections, but some of the reports in *The Toronto Star* and the *Globe and Mail* have been covering topics on global warming and climate change since as early as the late 1980s. Although a limited number of reports have been published some of them provided significant details of the science of climate change. One such report was published by *The Toronto Star* with the following caption: "We will be warm as toast 50 years down the road" (*The Toronto Star*, 3 February 1988). The report attributed projected warming to greenhouse effect, suggesting that Toronto had already been warming for some time. This particular report was based on a computer simulation done exclusively for *The Saturday Star* by federal weather experts.

Based on the Environment Canada simulations, the *Toronto Star* (3 February 1988) reported that within 50 years—and maybe even 30—climate will be so much milder that:

- Ice in the harbor and snow on the roads in mid-winter will be almost unheard of.
- Home heating costs will be down 30 to 45 per cent.
- A day when the temperature dips to minus 16°C (3°F), which we see now an average of three times every January, can be expected only once every five years.
- A day when the temperature goes above 31°C (88°F), which we see now an average of three times every July, will come an average of 12 times every July.
- The average temperature, day and night, year-round, will go up between 4 and 5 degrees Celsius.

Simulated hydrological impacts of higher air temperatures were also reported (*The Toronto Star*, 3 February 1988):

- Rainfall will stay roughly the same, but the higher temperatures will make so much more water evaporate that the land will be dry. Unless farmers change to crops more tolerant of this problem, agriculture in Ontario stands to lose $100 million a year.

- Because so much more water will evaporate, less will reach rivers to run into the lakes. That, combined with evaporation from the lakes themselves, will lower lake water levels by close to 1 metre (3.2 feet). Harbors such as Toronto's may be able to have shipping year-round.

As it was one of the earlier reports (*The Toronto Star*, 3 February 1988), it also provided a review of the greenhouse effect theory:

- The so-called "greenhouse effect" is nothing new for our planet. Research shows it started billions of years ago and without it, the life we know could not have flourished.
- It's caused by a number of gases found in tiny concentrations in the atmosphere. Carbon dioxide accounts for most of these. But almost as potent today, and even more worrisome in the long-term, are methane, nitrous oxide, low-altitude ozone, and man-made chlorofluorocarbons.
- All these gases let radiation from the sun reach the Earth's surface and warm the ground. But like a glass greenhouse roof, they make it difficult for heat to radiate back upward to space.
- For the past several million years, this natural greenhouse has kept most of the surface within a temperature range that suits plants and animals.
- Before humans developed agriculture and industry, the carbon dioxide concentration in the atmosphere stayed within a specific range for eons. But in 1957, scientists noticed the amount was going up. And in the 30 years since then it has gone up 10 per cent.
- As it goes up, the world gets warmer. And the carbon dioxide level is rising now at an alarming 4 per cent a decade, meaning the change is not only continuing, but speeding up.
- Factories do it with their chimneys, cars do it with their exhausts, homes do it with their furnaces. They all burn fuels that include carbon and give off gases that include carbon dioxide.

The *Globe and Mail* (18 December 1995) reported on anthropogenic (human-induced) global warming as the contributing factor for increasing air temperatures:

"Research shows Earth is now the warmest it has been in 600 years . . . Evidence suggests that man's folly, not the wrath of God, contributed to the weather chaos . . . The strange weather is consistent with computer models simulating the effects of global warming, thought to be caused mainly by the burning of fossil fuels" [said an Environment Canada scientist].

Anthropogenic global warming was expanded in a 1998 report by *The Toronto Star* (9 December 1998).

"The warming of the Earth in this century is without precedent in at least 1,200 years and cannot be fully explained by any known combination of natural forces, a top U.S. federal government climatologist says".

"New scientific findings . . . appeared to simultaneously add clarity and confusion to the debate. While some researchers reported strong signals of human-induced warming in the past century, other scientists acknowledged enormous uncertainties that complicate the task of forecasting climate change in the future".

"New findings add clarity—and confusion—to debate. One of the speakers, James Hansen, director of NASA's Goddard Institute for Space Studies, argues that scientists know too little about the complexities of climate, such as changes in cloud cover, to make accurate predictions. Hansen, who told a congressional panel in 1988 that the greenhouse effect "is here", caused a stir a month ago when he wrote about those uncertainties in a prominent journal, *The Proceedings of the National Academy of Sciences*. The forcings that drive long-term climate change are not known with an accuracy sufficient to define future climate change" [wrote Hansen].

Concluding comments

Differences in heat wave thresholds

This study has demonstrated that comparing heat waves in Delhi, India with that in Toronto, Canada is truly like a comparison between apples and oranges. The differences are more striking than similarities. The most obvious difference is in the threshold temperature for the definition of heat waves. Whereas 40°C (104°F) is considered as the beginning of a heat wave in Delhi, the comparable value for Toronto is 30°C (86°F), resulting in a threshold difference of 10°C (18°F). This difference is attributable largely to basic geographical contrasts between the climates of Delhi and Toronto. Being situated in the tropical/subtropical latitude, Delhi has a much warmer climate than a milder mid-latitude climate of Toronto. In general, Delhi experiences truly hot summer temperatures when 30°C is considered as normal temperatures. Our analyses of both official temperature records and newspaper reports indicate that Delhi experienced about 223 episodes of heat waves between 1978 and 1999. Each of these episodes consisted of one day to several days when the daily maximum temperatures exceeded 40°C. In comparison, Toronto experienced about 249 days of heat waves of different durations (i.e. much fewer episodes) during a period of six decades, i.e., 1938-1999, when the daily maximum temperatures exceeded 30°C. Thus, the average rates of heat waves in Delhi amounted to 10 episodes (greater than 10 days) per year; comparable data for Toronto was 4.2 days per year. It should be stressed here that although the Canadian threshold for heat waves is 30°C, often it is associated with humidex values reaching 40 or higher. The *humidex*, an index combining air temperatures with relative humidity values, is a Canadian innovation (by Environment Canada meteorologists). Although any equivalent of humidex was not used by the Indian Meteorology Department (IMD), our estimates based on the Canadian humidex table (chart) indicate that most often the Delhi threshold of 40°C was equivalent to "dangerous humidex" values of 50 or higher if relative humidity values were included. For example, if we interpolated (applied) data from the Canadian humidex chart, the maximum humidex values during some of the days of the second heat wave of May 2015 (18-31 May, see Table 7.5) ranged between 50 and

56 ("dangerous humidex") at record temperatures of 40°C to 45.8°C with relative humidity ranging between 20% and 40%.

Differences in fatality rates

According to the Canadian humidex chart, humidex values approaching 50 or higher are likely to result in severe dehydration and heat strokes. For this reason the rates of heat-related deaths (especially due to heat strokes) were significantly higher in Delhi than in Toronto. According to official sources cited by Murari and others (2015), heat-related death rates in Delhi were approximately 241 per year between 1978 and 1999 (a total of 5300 deaths in 22 years), compared to 120 per year in Toronto between 1938 and 1999 (Pengelly and others 2007). Even these significantly lower death rates were greatly under-reported by the three Canadian newspapers analyzed for this study, perhaps because of the technical difficulties of reporting less-than-straightforward mortality data (as we have explained this in the main text earlier).

Differences in newspaper discourse

One of the striking characteristics of newspaper reports on heat waves in Delhi and Toronto was that the Delhi heat waves had attracted widespread international media attention. As data in Table 7.6 indicate, a large number of reports were published in international newspapers, many in some of the leading US newspapers. In contrast, nearly 80% of the reports on heat waves in Toronto were published in three leading Canadian newspapers, namely, *The Toronto Star*, the *Globe and Mail*, and the *National Post*.

More importantly, there were significant differences in the nature of newspaper reporting on heat wave issues in two different environmental and cultural contexts. In case of Delhi, there was an emphasis on extreme temperature records of Delhi. Human miseries due to heat waves, especially the suffering of the poor, were often expressed in symbolic language using different types of metaphors. We found very little evidence of such symbolic language in reports on heat waves in Toronto, as most of the newspaper reporting seemed to be mundane. There were also significant differences in coping strategies. In case of Delhi, there were frequent references to power blackout, some with reports on air-conditioning problems. The latter was a major issue in many Toronto reports. Absence of workers in offices and factories, school closures, and increased sale of cold drinks and bottled water (including sale of ice-cold water by a poor boy) were some of the common coping strategies in Delhi. There was a report on a large number of people immersing in river water near Delhi—an unlikely coping strategy for Canadians—to get some relief from heat but there were few reports of swimming pools in Delhi. In contrast, there were several reports of people over-crowding swimming pools in Toronto. Many people took shelter in several formal cooling centres in Toronto as well as in air-conditioned shopping malls and movie theatres. Like Delhi, there were reports on increased sale of cold drinks and ice cream but increased sale of beer was largely a Canadian phenomenon.

Planning implications of heat waves

Current heat waves may be considered as an advance warning for significantly warmer days ahead, as most of the climate models project. Referring to the Indian success story of the Hyderabad model of 'Heat Action Plan' (that included cooling centres in public buildings), some of the newspapers reported that similar public health initiatives were underway in Delhi. Yet, it was not clear if such initiatives were sponsored more formally by either the city government or by any other levels of governments. In newspaper discourse there were many references to different types of cooling centres in Toronto, including four formal cooling centres and other informal centres, such as shopping malls, but their numbers seemed to be woefully inadequate compared to the large numbers of affected population of the metropolis. Most of the political discussion around climate change is on reductions of carbon dioxide emissions. Certainly, this is an appropriate strategy for addressing the root causes of climate change but urgent action is needed for alleviating current human sufferings from recurring heat waves, not only in Delhi and Toronto but also in many other parts of the world. Among potential adaptations to heat waves at least two measures seem to be feasible:

- Building an extensive network of cooling centres, especially in major urban centres
- Innovations in cooling technology for improving air conditioning

While the first measure would require major planning and financial investments by different levels of governments, scientific advances in cooling technology have already been reported in a recent article in *Scientific American* with a caption: "New Technologies Aim to Save Energy—and Lives—with Better Air-Conditioning" (Biello 2013). The main message of this chapter is summarized below:

> "Conventional air-conditioners employ refrigerants such as chlorofluorocarbons to absorb heat from the room to be cooled. That heat is then expelled outside, requiring electrically powered pumps and compressors. One idea to conserve energy is to replace coolant fluids and gases—which are often super-powered greenhouse gases capable of trapping more than 1,000 times more heat than CO_2—with solid materials, such as bismuth telluride. A new device from Sheetak, developed in part with ARPA-E funding, uses electricity to change a thermoelectric solid to absorb heat, and could lead to cheaper air-conditioners or refrigerators".

References

Biello, D. 2013. "New Technologies Aim to Save Energy—and Lives—with Better Air-Conditioning". *Scientific American*: 6 September 2013.
http://www.scientificamerican.com/article/technology-improvements-save-energy . . . retrieved on 1 August 2015.
Hannigan, J.A. 1995. *Environmental Sociology: A Social Constructionist Perspective*, 1st edition. London: Routledge.
Hannigan, J.A. 2014. *Environmental Sociology*, 3rd edition. London: Routledge.
Murari, K.K., Ghosh, S., Patwardhan, A., Daly, E. and Salvi, K. 2015. "Intensification of Future Severe Heat Waves in India and their Effect on Heat Stress and Mortality". *Regional Environmental Change* 15 (4): 569-579.
Pengelly, I.D., Campbell, M.E., Cheng, C.S., Fu, C., Gingrich, S. and Macfarlane, R. 2007. "Anatomy of Heat Waves and Mortality in

Toronto: Lessons for Public Health Protection". *Canadian Journal of Public Health* 98 (5): 364-368.

Rashid, H. 2011. "Media Framing of Public Discourse on Climate Change and Sea Level Rise: Social Amplification of Global Warming vs. Climate Justice for Global Warming Impacts". In *Climate Change and Growth in* Asia, edited by M. Hossain and E. Selvanathan, 232-260. Cheltenham, UK: Edward Elgar.

Smoyer-Tomic, K., Kuhn, R. and Hudson, A. 2003. "Heat Wave Hazards: An Overview of Heat Wave Impacts in Canada". *Natural Hazards* 28: 463-485.

Table 7.1 Monthly Temperature and Rainfall at Delhi (Palam)

Month	1901-2000 Maximum daily temperature in °C (°F)	1956-2000 Maximum daily temperature in °C (°F)	1901-2000 Minimum daily temperature in °C (°F)	1956-2000 Monthly rainfall in mm (inches)
January	20.8 (69.44)	20.8 (69.44)	7.8 (46.04)	18.9 (0.74)
February	23.7 (74.66)	23.9 (75.02)	10.3 (50.54)	16.6 (0.65)
March	29.6 (85.28)	30 (86)	15.4 (59.72)	10.8 (0.43)
April	36 (96.6)	36.9 (98.42)	21.5 (70.7)	30.4 (1.20)
May	39.8 (103.64)	40.5 (104.9)	26.2 (79.16)	29 (1.14)
June	39.4 (102.92)	40.3 (104.54)	28.3 (82.94)	54.3 (2.14)
July	35.2 (95.36)	35.4 (95.72)	27 (80.6)	216.8 (8.54)
August	33.6 (92.48)	33.7 (92.66)	26.2 (79.16)	247.6 (9.75)
September	34 (93.2)	34.2(93.56)	24.7 (76.46)	133.8 (5.27)
October	32.9 (91.22)	33.3 (91.94)	19.4 (66.92)	15.4 (0.61)
November	28 (82.4)	28.3 (82.94)	12.8 (55.04)	6.6 (0.26)
December	22.7 (72.86)	22.7 (72.86)	8.4 (47.12)	15.2 (0.60)
Annual	**31.31** (88.36)	**31.67** (89)	**19** (66.20)	**795.40 (31.30)**

Source: Prepared by the first author, based on original data from the Indian Meteorology Department. http://www.imd.gov.in/doc/climateimp.pdf . . . retrieved on 8 June 2015.

Table 7.2   Monthly Temperature and Rainfall at Toronto

Month	1981-2010 Maximum daily temperature in °C (°F)	1981-2010 Minimum daily temperature in °C (°F)	1981-2010 Monthly precipitation in mm (inches)	1981-2010 Monthly snowfall in cm (inches)
January	-1.5 (29.3)	-9.4 (15.08)	51.8 (2.08)	29.5 (11.61)
February	-0.3 (31.46)	-8.7 (16.34)	47.7 (1.88)	24 (9.45)
March	4.6 (40.28)	-4.5 (23.9)	49.8 (1.96)	17.7 (6.97)
April	12.2 (53.96)	1.9 (35.42)	68.5 (2.70)	4.5 (1.77)
May	18.8 (65.84)	7.4 (45.32)	74.3 (2.93)	0
June	24.2 (75.56)	13 (55.4)	71.5 (2.81)	0
July	27.1 (80.78)	15.8 (60.44)	75.7 (2.98)	0
August	26 (78.8)	15 (59)	78.1 (3.07)	0
September	21.6 (70.88)	10.8 (51.44)	74.5 (2.93)	0
October	14.3 (57.74)	4.6 (40.28)	61.1 (2.41)	0.4 (0.16)
November	7.6 (45.68)	-0.2 (31.64)	75.1 (2.96)	7.5 (2.95)
December	1.4 (34.52)	-5.8 (21.56)	57.9 (2.28)	24.9 (9.80)
Annual	**13 (55.4)**	**3.33 (37.00)**	**786 (30.94)**	**108.5 (42.72)**

Source: Prepared by the first author based on original data from Environment Canada. http://climate.weather.gc.ca/climate_normals/ results_1981_2010 . . . retrieved on 10 June 2015.

Table 7.3 Samples of Humidex Values

Temperature (°C)	Relative humidity (%)	Humidex	Discomfort level
22	100	31	Some discomfort
26	65	32	Some discomfort
26	100	39	Some discomfort
27	65	34	Some discomfort
27	95	40	Great discomfort
30	65	40	Great discomfort
30	100	48	Dangerous discomfort
32	75	46	Dangerous discomfort
32	100	52	Dangerous discomfort

Source: Environment Canada. Samples of data taken by the first author from the Environment Canada original table at https://ec.gc.ca/meteo-weather/default.asp/ . . . retrieved on 28 June 2015.

Table 7.4   Geography of Heat Waves in India, May 2015

State	Daily maximum temperature on 25 May: Exceeding 45°C (113°F) Areas of the state (percentages)*	Daily maximum temperature on 25 May: Between 40°C (104°F) and 45 °C (113 °F) Areas of the state (percentages)*
Rajasthan	Western, Northern, and Northeastern areas (33%)	Remaining areas (66%)
Punjab	Southern areas (10%)	Southern areas (50%)
Haryana	Southern areas (50%)	Northern area (about 50%)
Uttar Pradesh	Western and Southwestern areas (25%)	Northwestern and Southeastern areas (50%)
Madhya Pradesh	Northwestern and Northeastern areas (10%)	Remaining areas (90%)
Bihar	Southwestern areas (10%)	Central and Southwestern areas (50%)
Orissa	Northwestern areas (30%)	Remaining areas (70%)
Maharashtra	Northeastern areas (10%)	Remaining areas (90%)
Andhra Pradesh	Northeastern (coastal) areas (10%)	Remaining areas (90%)
Gujrat		Northern areas (50%)
West Bengal		Western areas (10%)

*   Source: All data in percentages may have substantial errors as these were estimated visually by the authors from a NOAA Climate Prediction Center map of India. Map source: http://www.slate.com/blogs . . . retrieved on 1 June 2015.

Table 7.5 Maximum Daily Temperature, Rainfall and Relative Humidity at Delhi, May 2015

Number of days with 40°C or higher temperatures	Maximum Temperature ranges	Rainfall (mm)	Relative Humidity at 2:00 pm (%)
First heat wave: 7 days (5-11 May)	40°C to 43.5°C	0	22-38
Second heat wave: 14 days (18-31 May)	40.6°C to 45.8°C	0	18-42

Source: Prepared by the first author based on original data from a website of the Indian Agricultural Research Station (located at Latitude 28 degree and 38 minute; Longitude 77 degree and 09 minute). http://iari.res.in . . . retrieved on 1 June 2015

Table 7.6 Samples of Newspaper Reports on Heat Waves in India: Delhi and Other Indian States, 1972-2015

Year (oldest to newest)	Newspaper	Daily maximum temperature at New Delhi in °C (°F)	Daily maximum temperature at another state in °C (°F)	Fatality
1972 (23 May)	*The New York Times*	37.8 (100)		300
1978 (22, 24 May)	*The Globe and Mail*, Toronto	45 (113)	47 (116.6)	150
1979 (22 May)	*The Globe and Mail*, Toronto		45.6 (114)	151
1983 (7 June)	*The Globe and Mail*, Toronto			67
1985 (3 June)	*The Globe and Mail*, Toronto		46 (114.8)	40
1988 (4 June)	*The New York Times*	45.6 (114)	48.9 (120)	450
1991 (11 June)	*The Guardian*, UK			200
1993 (29 May)	*The Atlanta Journal & Constitution*	44.4 (112)		
1994 (5 June)	*The New York Times*	44.4 (112)		130
1994 (14 June)	*The Washington Post*	46.1 (115)	49.4 (121)	400
1995 (12 June)	*The Scotsman*	45 (113)	47 (116.6)	167
1995 (16 June)	*San Jose Mercury News*	45.6 (114)	46.7 (116)	390
1998 (8 June)	*The New York Times*		50.6 (123)	2518
1999 (4 May)	*The Globe and Mail*, Toronto	44 (111.2)	47 (116.6)	110
2002 (21 May)	*The National Post*, Canada		40 (104)	781
2003 (5 June)	*Calgary Herald*, Canada		49 (120.2)	1200
2007 (13 June)	*Ottawa Citizen*, Canada	45 (113)	51 (123.8)	100
2010 (21 April)	*Right Vision News*	43.7 (110.7)	47.5 (117.5)	80
2013 (21 May)	*Political and Business Daily*, India	45.1 (113.2)	47.5 (117.5)	3
2014 (14 May)	*The Hindu*, India	41.4 (106.5)	38.5 (101.3)	
2015 (27 May)	*The Los Angeles Times*	45 (113)	47.8 (118)	1000
2015 (28 May)	*Siasat Daily*, India	50 (122)	45 (113)	1100
2015 (30 May)	*The Star*, South Africa		46 (114.8)	1800
2015 (9 June)	*China Daily*	42 (107.6)		2360

Source: Prepared by the first author based on the *LexisNexis® Academic* database for newspaper reports (877 reports) on heat waves in India, 1972-2015.

Table 7.7 Geography of Heat Waves in India: Interpreted from *The Times of India* **Reports, 2010-2015**

State (in order of frequencies of reports)	Frequency
Delhi (the city and also a union territory)	20
Rajasthan	13
Uttar Pradesh	13
Haryana	12
Punjab	12
Madhya Pradesh	5
Chattishgarh	4
Gujrat	4
Himachal	4
Jammu and Kashmir	4
Telangana	4
Andhra	3
Jharkhand	3
Bihar	2
Maharashtra	2
Orissa	2
Uttarkhand	2

Source: Prepared by the first author based on a sample of 25 reports of *The Times of India*.

Table 7.8 Maximum Daily Temperatures at Delhi and Other Indian States, 2010-2015: Reported in *The Times of India*

Year (number of reports)	Delhi: Maximum daily temperature in °C (°F) (range)	Other Indian states: Maximum daily temperature in °C (°F) (range)
2010 (8 reports)	41.6 to 47.6 (107 to 118)	45.1 to 48.1 (113 to119)
2011 (2 reports)	40.6 to 42.2 (105 to 108)	44.5 (112)
2012 (3 reports)	43.5 to 45.4 (110 to 114)	44.5 to 45 (112 to 113)
2013 (4 reports)	45.1 to 46.4 (113 to 116)	40.2 to 48.2 (104 to 119)
2014 (3 reports)	44 to 47.2 (111.2 to 117)	48.3 (119)
2015 (5 reports)	40.6 to 46.3 (105 to 115)	44.6 to 45 (112 to 113)

Source: Prepared by the first author based on *The Times of India* reports, 2010-2015

Table 7.9 Heat Waves in Toronto, 1938-1998

Maximum daily temperatures equal to or exceeding 30°C (86°F) for consecutive days	**Number of days** (for the entire period)
2 days	105
3 days	78
4 days	33
5 days or longer	33
Heat waves per year: 4.2 days	Total: 249 days

Source: Data obtained from Table II in Smoyer-Tomic and others (2003)

Table 7.10 Maximum Daily Temperatures at Toronto, 23-29 August 1948

August 1948	Maximum daily temperature in °C (°F)
23	30.6 (87)
24	35 (95)
25	38.3 (101)
26	36.7 (98)
27	35.6 (96)
28	37.8 (100)
29	32.2 (90)

Source: Prepared by the first author based on original data from Environment Canada. http://climate.weather.gc.ca/index_e.html . . . retrieved on 29 June 2015

Table 7.11 Samples of Humidex Values for Toronto, 17 July 2012

Hour (samples)	Air temperature (°C)	Relative humidity (%)	Humidex*
8 a.m.	29.1	55	36
9 a.m.	32.1	49	40
12 noon	34.8	44	43
1 p.m.	36.2	42	45
7 p.m.	32.1	53	41
8 p.m	27.2	68	35

* Number of hours with continuous humidex values of 40 or higher on 17 July 2012: 10 hours (9 a.m. to 7 p.m.).

Source: Prepared by the first author based on original data from the following website: http://climate.weather.gc.ca/ClimateData/hourlydata_e.html? . . . retrieved on 29 June 2015.

Table 7.12 Heat Waves in Toronto, July 2011

Heat wave sequence	Heat wave duration	Dates	Temperature ranges, °C (°F)
First heat alert	1 day	3 July	30 (86°F)
First heat wave	2 days	5-6 July	30.1 to 30.6 (86.18 to 87)
Second heat wave	3 days	10-12 July	31.8 to 33.5 (89.24 to 92.3)
Third heat wave	3 days	16-18 July	31.2 to 35.2 (88.16 to 95.36)
Fourth heat wave	4 days	20-23 July	32.1 to 37.9 (88.16 to 100.22)
Fifth heat wave	2 days	30-31 July	30.9 to 32.8 (87.62 to 91)

Source: Prepared by the first author based on original data from Environment Canada. http://climate.weather.gc.ca/ClimateData/dailydata_e.html? . . . retrieved on 29 June 2015.

Table 7.13 Hourly Temperatures and Humidex Values at Toronto, 21 July 2011

21 July 2011: Hour	Air temperature (°C)	Relative humidity (%)	Humidex
8 a.m.	30.4	69	42
9 a.m.	32.6	61	44
10 a.m.	33.6	58	45
11 a.m.	34.9	54	46
12 noon	36.1	51	48
1 p.m.	36.6	49	48
2 p.m.	37.1	47	48
3 p.m.	37.7	46	48
4 p.m.	37.5	43	48
5 p.m.	36.6	45	47
6 p.m.	35.6	45	45
7 p.m.	34.5	50	44
8 p.m.	33.1	56	43
9 p.m.	31.2	62	42

Source: Prepared by the first author based on original data from Environment Canada. http://climate.weather.gc.ca/ClimateData/hourlydata_e.html? . . . retrieved on 29 June 2015.

Table 7.14 Major Themes in Newspaper Reports on Heat Waves in Toronto, 1981-2014

Major themes → Date: Newspaper	HWD (%)*	HLI (%)	POP (%)	OTP (%)	COP (%)	CCI (%)
1981 July 10: *Globe & Mail*		100				
1987 June 17: *Globe & Mail*	64		10	27	9	5
1987 July 15: *Toronto Star*			100		26	
1988 Feb 13: *Toronto Star*	7				6	76
1988 July 9: *Globe & Mail*	11		5	57	27	
1988 July 10: *Toronto Star*		43	28		25	
1988 July 11: *Toronto Star*	10	21			69	
1988 July 12: *Toronto Star*				100		
1988 July 13: *Toronto Star*	6	100				
1989 June 26: *Toronto Star*	100					
1989 July 11: *Toronto Star*	58				33	
1993 Aug 27: *Toronto Star*	45	4			52	
1994 June 18: *Toronto Star*	15	43			55	
1995 June 20: *Toronto Star*	24	14	12	69	40	
1995 Dec 12: *Globe & Mail*						100
1998 Dec 9: *Toronto Star*	7					93
1999 July 14: *Toronto Star*		21			78	
2003 June 25: *Toronto Star*	36	13	9		40	
2005 July 20: *Toronto Star*	20	51	17		11	
2006 June 9: *Toronto Star*					25	74
2006 Aug 2: *Globe & Mail*	10		15		44	33
2006 Aug 2: *National Post*	100					
2012 June 21: *Toronto Star*	22		53		13	
2013 June 24: *Globe & Mail*	57	20			23	
2014 Aug 25: *Toronto Star* (*digital version: star.com*)	52				48	
Total numbers of words: **12,041**	**20**	**13**	**10**	**11**	**26**	**20**

*   % indicates percent of the total number of words in a report. The total may exceed 100% because of overlapping concepts in the text. Numbers of words were counted for the following issues: (a) HWD: Heat wave details, (b) HLI: Health issues, (c) POP: Power problems, (D) OTP: Other problems, (e) COP: Coping strategies, (f) CCI: Climate change implicated.

Source: Prepared by the first author based on content analysis of a sample of 25 newspaper reports drawn blindly (using a random number table) from an initial sample of 121 web-based reports (drawn from *LexisNexis® Academic* database for 991 newspaper reports).

Table 7.15 Comparison of Environment Canada Data on Maximum Daily Temperatures with Those Reported in Newspapers

Newspaper: Date	Newspaper data (°C)	Environment Canada data (°C)
*TS: 1988 August 3	34.7	34.7
TS: 1989 July 10	33	32.1
TS: 1993 August 26	31	31.5
TS: 1994 June 17	32.59	33.3
TS: 1995 June 19	35.6	35.5
TS: 1999 July 4	34.3	34.6
TS: 2003 June 24	33	32.9
TS: 2012 June 21	34	34.4
*GAM: 2006 August 1	36.6	36.6

*   TS: *The Toronto Star*; GAM: *The Globe and Mail* (Toronto)

Source: Prepared by the first author based on newspaper-based data and Environment Canada data on maximum daily temperatures.

http://climate.weather.gc.ca/ClimateData/dailydata_e.html? . . . retrieved on 12 July 2015.

Chapter 8

Discourse Analysis of Newspaper Reports on Climate Change Politics in USA

CLIMATE CHANGE CONTROVERSIES

Starting from the last two decades of the 20th century, public discourse on climate change has emerged as one of the most popular yet contested environmental issues of the 21st century. Much of the climate change controversies have resulted from the global warming hypothesis. Unlike objective climatic data, such as air temperature, air pressure, precipitation and others, global warming is an assumption that attributes current increasing trends in global average air temperatures to increasing emissions of greenhouse gases. A large body of scientific evidence suggests that urban-industrial emissions of carbon dioxide and other greenhouse gases—often called anthropogenic greenhouse gases—have been mainly responsible for the current global warming trends. Citing natural cycles of many past global warming episodes, some during past several thousand years and others during geological periods spanning millions of years, a small group of scientists (the so-called climate change sceptics) has challenged the anthropogenic global warming hypothesis, suggesting that the warming trends might be attributable to several natural geophysical dynamics, such as sunspot activities, changing angles of earth's rotational axis, and reversal of magnetic poles. The skeptics argue that many of the current climate change impacts (i.e.

impacts of global warming on different characteristics of the climate), such as an enhanced hydrologic cycle (i.e. a more rainy climate with resulting increases in higher-magnitude floods), greater storminess (especially greater frequencies of high-magnitude hurricanes/cyclones) and increasing events of droughts, are difficult to correlate directly to global warming because of the compounding feedback effects of a warming atmosphere and its hydrologic cycle. Further, climate models have also been a source of climate change controversies, despite their enormous contributions to the advancing science of climate change. Climate models are assumptions and the outputs of such models are contingent upon the nature of the data inputs. In public discourse on climate change, model projections for future climate change scenarios have often been misinterpreted by non-specialists and sometimes grossly misused by interested parties for rhetorical purposes. The central problem with political rhetoric on climate change is that climate change impacts as well as their mitigation and adaptation measures have both economic and political implications as these are likely to affect current livelihood activities of many voters either adversely or positively. In short, climate change is an extremely complex environmental problem that has polarized the voting public in the United States. Although support for climate change policies may vary within memberships of each of the political parties, the majority of the Democrats seem to favor different versions of climate change policies whereas most of the Republicans are largely opposed to such policies. Coincidentally, many in the Republican leadership and their lobbyists who oppose climate change policies rely heavily on the science espoused by climate change sceptics.

GLOBAL WARMING IN USA

Greenhouse gas emissions

Global warming may be defined as "the idea that increased greenhouse gases cause the Earth's temperature to rise globally" (Houghton 2004, 335). Contrary to the current debate about anthropogenic global warming, objective scientific data on greenhouse gas emissions and average air temperatures in USA indicate that both of them have been increasing for several decades (Tables 8.1 and 8.3).

Data in Table 8.1 represent emissions by at least 26 anthropogenic greenhouse gases (EPA 2016, see Executive Summary [ES], Figure ES-5, page ES-9). The term *anthropogenic greenhouse gases* literally means emissions and removals (alternately called *sinks*) of greenhouse gases that are direct results of human activities. This does not necessarily imply that they are entirely human-induced. According to a more comprehensive definition, it includes emissions and removals that are "a direct result of human activities or are the result of natural processes that have been affected by human activities" (EPA 2016, ES-1). All data in Table 8.1 are expressed in CO_2 equivalents because each of the greenhouse gases has a different *global warming potential* (GWP), which depends on a process called *radiative forcing*, i.e. the relative efficiency of a given gas to absorb solar and earth radiation. Compared to the GWP value of 1 for carbon dioxide, methane for example has a GWP of 25; i.e. it has a warming potential of 25 times higher than that of carbon dioxide. Nitrous oxide (laughing gas) has a GWP of 298 (EPA 2016, page ES-1). Some of the other greenhouse gases have much higher GWP—some as high as 17,200—but fortunately these are trace elements, i.e. a minute quantity occurs in our atmosphere (EPA 2016, Figure ES-5, page ES-9).

Between 1990 and 2014 greenhouse gas emissions in USA increased by 7.7%, i.e. at an average growth rate of 0.3% per year. The peak emissions occurred in 2005 with a record emission of 7,429 million metric tons (or 7.4 Gt [gigatonnes]). The main sources of different types of greenhouse gases are listed in Table 8.2. Among 26 greenhouse gases listed in EPA (2016), carbon dioxide, methane and nitrous oxide accounted for 97% of total greenhouse gas emissions (Table 8.2). Growth rates of greenhouse gas emissions by different gases have been uneven between 1990 and 2014: whereas carbon dioxide emissions increased by 8.6%, nitrous oxide increased by only 0.5% and methane emissions declined by 5% (EPA 2016). The latter has resulted largely from improvements in "decomposition of wastes in landfills, enteric fermentation associated with domestic livestock, and natural gas systems" (EPA 2016, ES 7). The bulk of the carbon dioxide emissions (94%) originated from fossil fuel combustion (Table 8.2). Five main economic sectors accounted for 99% of fossil fuel combustion. The most striking data in Table 8.2 is that nearly three-quarters of carbon dioxide emissions originated from electricity generators and transportation (mainly cars, trains, ships, and airplanes). The industrial and commercial sectors accounted for only one-fifth of the carbon dioxide emissions. The residential sector was a relatively small contributor (7%).

Average temperatures in USA

Increased concentration of carbon dioxide in the atmosphere is the direct result of greenhouse gas emissions. For centuries, global atmospheric carbon dioxide level had never been above 300 ppm (parts per million) (NASA 2016). Starting from the 1950s, carbon dioxide levels have been increasing sharply from about 320 ppm to 400 ppm, as of 2016 (NASA 2016). Since carbon dioxide absorbs both shortwave solar radiation and longwave earth radiation such absorptions of radiation energy by the atmosphere have increased global average air temperatures. Consistent with the global trend, average air temperatures within the territories of USA have also increased correspondingly.

Annual temperatures of a given station may be computed by averaging the monthly temperatures for 12 months (January to December). The monthly temperatures, in turn, are averages of daily temperatures in a given month (such as 31 days in January and 28/29 days in February). In other words, annual temperatures are essentially averages of daily temperatures for 365 days in a given year. The *average annual temperatures* (often called *mean annual temperatures*) are computed by averaging annual temperatures at a given station for a given period, such as a decade or a century or any other time interval. For example, the average annual temperature of 51.54 °F (10.86 °C) for 1901-1910 represents an average of records for 3,650 days (i.e. 365 days x 10 years). The NOAA (2016) database which is readily available from its website (see reference at the bottom of Table 8.3) provide instant averages for any time periods for annual temperatures of 48 contiguous states from 1895-2016.

The average annual temperatures in USA also integrate spatial variations in temperatures within the U.S. mainland (48 contiguous states). Air temperatures differ significantly between the interior of the country (continental climate) and the coastal belts (marine climate) and also between lower latitudes (such as Texas, Louisiana, Florida, to name a few) and midlatitudes (such as Minnesota, Iowa, Wisconsin and others). However, average annual temperatures have increased in every region of the country, irrespective of differences in regional climates. Based on a National Climatic Assessment (NCA 2014) map, some of the highlights of changes in average annual temperatures within the United States mainland (48 contiguous states) are as follows:

Major increases (1°F to 2 °F) in some of the states (in the entire area of a given state or in only parts of a given state): all of the northeastern states, most of the Midwestern states, some of the southwestern states, some of the interior states, southern tip of Florida.

Moderate decreases (-0.1°F to -1°F): Alabama, Mississippi, Arkansas, Oklahoma, and Kentucky.

Moderate increases: (0.1°F to 0.5 °F and some up to 1°F): The remaining states and parts of the above states (under major increases).

In short, the rates of temperature increases (within the range of 0 to 2 °F) have been highly uneven throughout the United States. For example, some parts of a given state warmed up to 2°F, while the average annual temperature increased by about 1 °F in other parts. Data on average annual temperatures in USA, as presented in Table 7.3, are spatially significant in the sense that they represent averages of data for thousands of stations throughout different regions of the country. Similarly, they are also temporally significant because they represent averages of large numbers of data for many years.

The average annual temperatures in USA increased by 0.17°F per decade (0.1°C per decade) between 1901 and 2010. Within this period temperatures declined during three decades: 1911-1920, 1941-1950 and 1961-1970.

Global warming is also evident in average seasonal temperatures. All seasons have become warmer. Winters in particular have warmed up more than other seasons. January temperatures, representing the peak winter season have increased by about 0.71°F in the second half of the record (1961-2015). Temperature increases in spring (typified by April), summer (July) and fall (October) have been more modest: by 0.58°F, 0.32°F and 0.16°F, respectively.

METHODS OF DISCOURSE ANALYSIS

Main objective of this study

Digital copies of newspaper reports provide a readily available source of information on climate change politics in the United States (and in many other countries of the world). Based on a sample of U.S. newspaper reports and a method of discourse analysis of those reports, the main objective of this study is to analyze political discourse on climate change in USA with the goal of gaining insights into some of the major climate change issues that have recently been driving climate change politics in USA.

Definition of discourse analysis

Discourse analysis is "a method of analyzing the structure of texts or utterances longer than one sentence; taking into account both their linguistic content and their sociolinguistic context" (Barber 1998, 399). More specifically, it deals with the analysis of a body of statements—both written and spoken—concerning a specific subject matter, especially "as typified by recurring terms and concepts" (Barber 1998, 399). By delineating different concepts in a body of text, such as newspaper reports, discourse analysis can help in identifying competing climate change policy options.

Sampling method

We used *LexisNexis® Academic*, an electronic database (subscribed by Murphy Library of the University of Wisconsin-La Crosse) as the primary tool for retrieving relevant newspaper reports on climate change politics in USA. We initiated our search by using the key phrase "climate change politics in USA". This provided a list of 1,000 reports published in newspapers, newswires, press releases, magazines, and others (as of 15 April 2016). By far, the largest percentages of these documents (93.5% or 935 out of 1,000) were international and U.S. newspaper reports. Table 7.8 lists leading newspapers ranked by the *LexisNexis* database, based on the numbers of reports on climate change politics in each of the newspapers.

Our objective was to sample only U.S. newspapers. On a closer scrutiny of several reports in some of the U.S. newspapers listed in Table 7.5 we found that many of them did not cover substantive aspects of either climate change or of climate change politics. Some of the reports were misclassified by the word "climate" or "climate change', which appeared sparingly throughout these reports, often in an entirely different context (such as "political climate") in a single sentence or in couple of sentences without any substantive coverage of the topic. To represent a sample of substantive reports on climate change politics in USA, we browsed through a large number of reports and selected the following three newspapers: *The New York Times, The Christian Science Monitor*, and *USA Today*. To conform to our original goal of preparing a popular science text for the general audience, we did not use a rigorous sampling method, such as sampling based on a computer-generated random number table. However, the selection of these newspapers was not entirely arbitrary; it was based largely on the depth of coverage of news on climate change politics in U.S.A. and partly on their name recognition and circulation data. The *USA Today* had by far the largest circulation: about 4.139 million in 2014 (http://www.usatoday.com/ . . . retrieved on 7 April 2016). *The Christian Science Monitor* transformed its print media into digital publication as of 2009. As of 2012, it had a digital traffic of about 8-10 million (http://www.csmonitor.com/ . . . retrieved on 7 April 2016). *The New York Times* had a circulation of about 2.178 million (Monday to Friday editions) and 2.624 million (Sunday edition), as of 2015 (http://www.nytco.com/ . . . retrieved on 7 April 2016).

Although we started our search with 1,000 reports in the final analysis we ended up with only 61 reports in three US newspapers that contained substantive coverage of climate change issues (we did not review international newspapers). Out of them, 47 reports were written by professional journalists. In most of these reports names of specific journalists who had written them were indicated under "by-line". Some of the reports were published as political news or without any categories and without specifying any authors. Several reports were editorials or op-ed columns. Table 8.7 summarizes the main categories of newspaper discourse in 47 reports. The remaining 14 reports contained 35 letters to the editors, which represented direct inputs on climate change from the readers (Table 8.8). In all, 61 reports had a total word count of 39,621 with the following breakdown: 47 reports contained 34,975 words (87% of the total text) and 35 letters (in 14 reports) contained 4,646 words (13%). Among the three newspapers *The New York Times* (NYT) contained more comprehensive coverage of climate change issues than either *The Christian Science Monitor* (CSM) or *USA Today* (UST). This is reflected in different types of descriptive statistics (data) on lengths of respective newspaper reports on climate change. Thus, the average length of the NYT reports was 1,239 words, compared to 710 words for CSM reports and 707 words for UST reports. The median length of NYT reports (1,056 words) was also significantly higher than that of the remaining two newspapers. However, the single longest report (1,468 words) was published in *USA Today* (Table 7.6).

Classification of texts

Discourse analysis is a type of content analysis but we did not use any software for content analysis or discourse analysis. Instead, we used a manual method of classifying the text into a number of segments, each related to one of the climate change themes. The method involved several successive steps. **First**, to begin with each report was saved as a Microsoft Word document. Then, we browsed through several news reports and identified the following five recurring themes: (a) global warming hypothesis, including discourse on climate change sceptics/deniers, (b) climate change impacts, including sea level rise, (c) mitigation and adaptation measures, including reductions of greenhouse gas emissions and adoption of alternative or renewable energy sources, (d) climate change politics and policies, and (e) others that did not address any aspect of climate change issues. In media research delineation of such recurring themes is similar to *frame analysis*. Frames or news angles may be defined as "organizing devices that help both the journalist and the public make sense of issues and events and thereby inject them with meaning" (Hannigan 1995, 61). The literature on frame analysis is extensive. A peer-reviewed journal article by Fletcher (2009) dealing with political discourse on climate change policies during the Bush administration provides an example of the application of frame analysis as a part of the discourse theory.

Second, each of the selected news reports was highlighted and the total numbers of words in the report were recorded in an Excel spreadsheet. It may be noted here that as a part of its default function a Microsoft Word document displays its total word counts at the bottom of the document. Further, this elegant program provides instant word counts for any highlighted segment of a document. **Third**, following a thorough review of a given report its entire text was classified into one or more segments, each corresponding to one of the above five categories. For the purpose of cross-checking, each segment was labelled with one of the sub-headings, i.e. one of the five climate change-related themes. **Fourth**, word counts were obtained for each of the news segments of the text by highlighting a given segment. The resulting data for word counts were entered into the spreadsheet. **Finally**, to assess relative depths of news coverage on different aspects of climate change, space devoted to each of the classified news segments (i.e. numbers of words) was converted into percentages of total space in all reports (samples) of a given newspaper (Table 8.7). Data obtained in this manner provided a basis for assessing news density on different aspects of climate change for answering the central question: "What news reports were driving the story lines on climate change politics in USA?"

RESULTS AND INTERPRETATIONS OF NEWSPAPER DISCOURSE

Global warming hypothesis

Basic data on global warming

To put it simply, global warming is an increase in average global surface temperature. The projections for global warming have been provided at great lengths in successive IPCC Assessment Reports. Often it is difficult for newspaper readers (general audience) to keep track of a complex set of data on temperature projections under different scenarios. Newspapers play an important role in simplifying necessary data. For example, *USA Today* (2 March 2009) cited the 2007 IPCC Report providing its main projection that the average global surface temperature was expected to rise by 3.3°C (6°F) by 2100. Conforming to popular perceptions, the current and projected global warming is being fuelled mainly by anthropogenic greenhouse gases, particularly carbon dioxide emissions. *The Christian Science Monitor* (24 May 2007) reported that carbon dioxide emissions had been "spiraling upward faster than earlier predicted". In particular, China has emerged as a major contributor of carbon dioxide emissions for at least two decades, as its emissions "increased from 2000 to 2004 at a rate that is over three times the rate during the 1990s" (*The Christian Science Monitor*, 24 May 2007). Despite overwhelming scientific evidences, many Republicans, who may be characterized as climate change cynics, have taken issues with human role in global warming. In contrast, some of the Democrats who support the anthropogenic global warming hypothesis often tend to attribute every major climatic disaster as the outcome of climate change. This may or may not be true but each assumption requires further research to establish a link between a given disaster and climate change. Most of the newspaper reports have presented this debate in an objective manner as succinctly as possible. As data in Table 8.7 indicate, approximately 9-12% (or an average of 10%) of the total words of 46 reports in three newspapers were devoted to the global warming hypothesis.

Bipartisan debate on global warming

The New York Times (NYT) was the leader in newspaper discourse on climate change debate. Its most comprehensive report on climate change debate was published in its 2003 August 5 edition with the following heading: "Politics reasserts itself in the debate over climate change and its hazards". This was the second longest report of our sample (1,239 words) (Table 7.6). It was written by Andrew C. Revkin, NYT staff reporter, who summed up the growing climate change debate with his opening simile:

> "Just as the global climate ebbs and surges, with droughts followed by deluges, so does the politically charged atmosphere that has long surrounded research pointing to potentially disruptive global warming" (Andrew C. Revkin, *The New York Times*, 5 August 2003).

One of the paradoxes of this debate was that it was triggered by a bipartisan initiative to curb greenhouse gas emissions, a bill jointly sponsored by Republican Senator John McCain (Arizona) and Democratic Senator Joseph I. Lieberman (Connecticut). The paradox was that the Democrat-Republican bipartisan divide on climate change was not necessarily monolithic. Several Republicans like McCain supported emission cut initiatives as a practical necessity (for example for pollution control). McCain seemed to have agreed with the assumption that "scientists who have called for action and those who say risks from warming are overblown agree that it has been many years since research on warming has been the subject of such a vigorous assault" (*The New York Times*, 5 August 2003).

Republican positions on climate change

Republican Senator James M. Inhofe (Oklahoma), Chairman of the U.S. Senate Committee on Environment and Public Works (2003-2007, 2015-current), led the charge against the emission curbs:

> "With all of the hysteria, all of the fear, all of the phony science, could it be that man-made global warming is the greatest hoax ever perpetrated on the American people? It sure sounds like it" (*The New York Times*, 5 August 2003).

Mr Inhofe was banking on the work of a "small core of researchers" who insisted that there was "no evidence for human-caused warming of any import". The report devoted at least 500 words on the scientific and academic merit of a peer-reviewed journal article by two reputed astrophysicists who concluded that projected impacts of man-made global warming were exaggerated. According to the Center for American Progress Action Fund, a left-leaning advocacy group in Washington, at least 58 percent of Republicans in Congress have denied a link between human activity and global warming (*The New York Times*, 25 October 2014). The 2016 Republican Presidential candidate, Donald Trump has taken even a more extreme position by characterizing "global warming a hoax" and further claiming that "the

Chinese fabricated climate change ('just a joke, he later said')" (*The New York Times*, 20 may 2016).

The Republican positions on human role in global warming have changed very little during the last two decades. Some of the high profile Republican leaders, such as Senator Marco Rubio and Governor Rick Scott of Florida, who had all along been questioning whether human activity had an effect on climate change, have recently taken a more ambiguous stand by stating that "I am not a scientist" (*The New York Times*, 25 October 2014). However, confronted with the impacts of sea level rise, some of the Gulf Coast Republican leaders have been forced to take a more pragmatic approach to climate change-related issues. For example, pointing out the impacts of sea level rise in the Florida Keys, George Neugent, a Republican county commissioner said that "while people might disagree about what to do about climate change, the effects of flooding and hurricanes were less ambiguous. Clearly rising tides are going to affect us" (*The New York Times*, 25 October 2014). Pragmatism of field evidence seemed to have influenced Mr Neugent's personal views on climate change. "I have to be very careful when I say some things, especially to the skeptics," Mr Neugent said, adding that "he avoided arguments about the science of climate change. It's not worth the effort or the time to prove what clearly is a factual situation. We are living with it" (*The New York Times*, 25 October 2014). Like Mr Neugent, a growing number of new generation Republicans, especially at state and local levels, have taken a pragmatic approach for recognizing the reality of climate change. "I do not think that we want to be the party that believes in dirty air," said Mr James Brainard, the Republican mayor of Carmel, Indiana. George P. Bush, son of Jeb Bush, perhaps typifies the new generation Republicans whose views on climate change have strayed dramatically from the party line. During primary contests in 2014 for the office of Texas Land Commissioner the younger Bush suggested that climate change was a serious threat to Texas, though he stopped short of attributing it to human activity. Referring to the vulnerability of the Gulf Coast to storms, he said that climate change-related issues like sea-level rise and coastal erosion "honestly keeps me up at night" (*The New York Times*, 31 August 2014).

Democratic positions on climate change

While the majority of Republicans question anthropogenic global warming, the Democratic positions range between cautious evidence-based statements to more passionate, often simplistic assumptions about climate change. Hillary Clinton's position is an example of the former: "The science of climate change is unforgiving, no matter what the deniers may say; sea levels are rising, ice caps are melting, storms, droughts and wildfires wrecking havoc" (*The New York Times*, 13 April 2015). President Obama has championed the climate change issues during his tenure. In the process, even he has often been simplistic, or maybe reductionist, in his approach. For example, based on his assertion that climate change was a "settled science", Obama said in a *New York Times* interview in June 2014 that "climate change shows up in weather disasters like hurricanes and droughts . . . those start multiplying" (*The Daily Oklahoman*, 14 June 2014, reproduced in the *USA Today* as Oklahoman Editorials). The main problem with this type of assumption is that it is not scientifically tenable to relate each and every weather disaster to climate change.

Journalists' role in communicating climate change issues

Some of the journalists who have been covering climate change news on a routine basis have expressed their frustrations with the intractable nature of the subject matter. To vent such frustrations and to demonstrate his basic point on climate change, Brad Knickerbocker, a *Christian Science Monitor* reporter recommended his readers a popular science book by Mark Hertsgaard: *Hot: Living Through the Next Fifty Years on Earth* (*The Christian Science Monitor*, 14 March 2011). In his report, Knickerbocker summarized the objective of the book: "Hot: Hertsgaard's rhetoric device—considering global warming with a regard for the future children—is good without being too smarmy" (*The Christian Science Monitor*, 14 March 2011). To demonstrate that climate change need not be an intractable issue if we could focus on its root causes, Knickerbocker cited a couple of sentences from Hertsgaard's book that capture the central theme of anthropogenic global warming hypothesis: "Because of the increase in greenhouse gases that began with the Industrial Revolution and accelerated in the 20th century with industrial agriculture, motor vehicles, and worldwide population increase . . . over the next fifty years, climate change will transform our world in ways we have already begun to imagine" (*The Christian Science Monitor*, 14 March 2011).

Television meteorologists' contributions

Because of their formal training in meteorology and climate science many television meteorologists, like print journalists, have developed expertise in explaining climate change in plain language. For example, in a *USA Today* report Bob Ryan, the lead meteorologist at NBC4, Washington, DC, provided some insights into the global warming hypothesis: "Today, we have a far greater understanding of everything from dynamic glacial changes in Greenland to ocean-atmosphere interactions (such as El Niño/La Niña) to rising sea levels, solar cycles and nearly every other physical process on earth than we had just three decades ago" (*USA Today*, 9 May 2009). Scientists may not be able to predict climate change precisely but Ryan's comparison of weather forecasts with model-based climate change forecasts is helpful for understanding projections on climate change. He wrote that "In day-to-day forecasts, the basic chemistry of the air and oceans does not change, and the prediction is on a small scale at one point in time" In contrast, "in a prediction of earth's temperature 50 to 100 years out, the critical chemistry of the air and ocean does change, as is happening with the rapid increase in CO_2 in the air and the rising acidity of the oceans". He went one step further by placing his confidence in long-range climate change: ". . . there is increasing confidence that major changes in regional and national climates will occur within a few generations" (*USA Today*, 9 May 2009).

Climate change impacts

Climate change impacts are changes in different characteristics of the climate (atmospheric environment) and earth surface environment due to global warming. Most of the newspaper reports on climate change impacts focused on current impacts of global warming. As data in Table 8.7 indicate, 6-8% of the text reported on different types of climate change impacts, such as sea level rise, coastal erosion, loss of sea ice cover in the Arctic Ocean, retreat of land glaciers, intensification of hydrological regimes, increases in droughts and forest fire and others.

Sea level rise

Sea level rise was reported as a dominant issue along both the Atlantic and the Gulf Coasts. A North Carolina state commission of experts estimated that the sea level could rise along its coasts by 39 inches, whereas along the southeastern Virginia coast (for example, along the low-lying Hampton Roads) the sea level "is likely to rise at least a foot in 30 years and five feet or more by the end of this century" (*The New York Times*, 7 July 2014). In Texas, the state's 367-mile Gulf Coast has become increasingly vulnerable to sea level rise, coastal erosion and storm surges that scientists say "are all exacerbated by climate change" (*The New York Times*, 31 August 2014). Another *New York Times* report (3 September 2015) described President Obama's historic visit to Arctic Alaska for field evidence (firsthand knowledge) on the nature of climate change impacts. Arctic Alaska, in particular, has been severely impacted by retreat of land glaciers, loss of sea ice cover and coastal erosion. This report touched on some of these issues citing a statement from the White House: "In Arctic Alaska, villages are being damaged by powerful storm surges, which, once held by sea ice, are battering the barrier islands . . . Alaskan Native traditions that have set the rhythm of life in Alaska for thousands of years are being upended by decreasing sea ice cover and changing seasonal patterns" (*The New York Times*, 3 September 2015). The report further indicated that President Obama was "the first sitting president to visit Arctic Alaska" to see firsthand the nature of impacts of sea level rise, particularly "coastal erosion in a small village called Kotzebue" (*The New York Times*, 3 September 2015). Among other impacts, loss of polar bear habitat due to depletion of sea ice cover has been one of the most adverse ecological impacts of climate change. In 2006 (10 February), *USA Today* reported that the federal officials were considering declaring the polar bear a threatened species because of "rising Arctic temperatures and melting the ice pack that's their home".

Glacier retreat

Another visible impact of climate change in Alaska is glacier retreat (due to melting). During his visit to Alaska Mr Obama went to the Kenai Fjord National Park (on 1 June 2015) and hiked to the Exit glacier—which has "receded more than a mile over the past 200 years, but has hastened considerably in recent decades—to bear witness to the change. He called the receding glacier 'a snapshot' of the impact of the planet's warming" (*The New York Times*, 3 September 2015).

Impacts on hydrologic cycle

Besides sea level rise and glacial retreat in the Arctic, climate change impacts have affected other regions of USA in different ways. The hydrologic cycles of different interior basins have been intensified significantly. For example, Commissioner Paula Brookes of Franklin, Ohio, which includes Columbus, the state capital, indicated that "there had been a 37 percent increase in flooding in the area since 1958, as heavy rains have overwhelmed aging drainage systems. The runoff from such rains has carried fertilizer into Lake Erie, contributing to algae crisis that forced Toledo, Ohio, to ban the use of tap water for several days in August" (*The New York Times*, 25 October 2014). Other impacts that have been reported include "declining water resources, drought, and wildfires. In the heavily forested Pacific Northwest, for example, scientists predict more trees will die because of insect infestation" (*The Christian Science Monitor*, 24 May 2007).

Impacts on forest fire

The 2007 wildfires in southern California had been attributed to global warming, which had "turned up the political and scientific heat about climate change's possible role in the conflagrations" (*The Christian Science Monitor*, 1 November 2007). Walter Oechel, a biology professor at San Diego State University, who had to be evacuate from his home, saw it in personal terms: "The fires we just experienced are some of the first effects we are feeling from climate change. We now have a very graphic representation of what many of us have been saying for a long time" (*The Christian Science Monitor*, 1 November 2007). Still the connection between global warming and forest fire is not necessarily straight-forward. Researchers at the University of California, Merced and the University of Arizona said in a statement: "At present the connection between global warming, Santa Ana winds [hot winds descending from mountains], and extremely low Southern California precipitation last winter are not known with sufficient certainty to conclusively link global warming with this disaster" (*The Christian Science Monitor*, 1 November 2007). The same statement continued: "Climate model projections suggest that with rising greenhouse-gas concentrations in the atmosphere, these phenomena will become increasingly likely in the future" (*The Christian Science Monitor*, 1 November 2007).

Questioning climate change link to hurricanes

When Hurricane Sandy struck the U.S. Atlantic coasts (in late October 2012) it was a category 1 storm (with 80 miles per hour winds) but it was the deadliest and most destructive storm of 2012 since it struck heavily populated areas of the northeast coast. New Jersey, in particular, sustained the greatest amounts of damage from both hurricane winds and storm surge flooding. New York was also affected significantly: "a swollen New York Bay overflowed into Manhattan, flooding subways, tunnels, and a major power substation" (*The Christian Science Monitor*, 13 November 2012). In the aftermath of the storm many climate change activists started attributing this destructive storm directly to global warming. For example, Frances Beinecke, president of the Natural Resources Defence Council blogged: "We have entered a new era . . . climate change is heating up our oceans and pumping hurricanes and other storms with extra energy, more moisture, and stronger winds. It is swelling our seas, so that storm surges are higher and cause more flooding. From Norfolk, Virginia to Boston, sea levels are rising four times as fast as the global average. Hurricane Sandy cut right along those swelling seas" (*The Christian Science Monitor*, 3 November 2012). The problem with this assumption was that "the connection between a single 'hybrid storm' like Sandy [which had multiple sources of energy from both mid-latitude and tropical low pressure systems] and Earth's generally agreed upon warming trend tied to industrial and motor vehicle greenhouse gas emissions" was not necessarily straight-forward (*The Christian Science Monitor*, 3 November 2012). The general public may not be aware of such scientific nuances. On the contrary, there is a widespread perception that there is a connection between global warming and hurricanes. Newspapers can play an important role by explaining the caveats, as *The Christian Science Monitor* (3 November 2012) made an attempt to do so in its report with the following caption: "Climate scientists caution against any direct connection between a hybrid storm like Sandy and Earth's warming trend".

Mitigation

Climate change mitigation refers to measures to reduce or prevent greenhouse gas emissions. This may be achieved through several measures: (a) political initiatives/management policies to minimize or prevent emissions, (b) use of renewable or alternative energy sources, (c) transformation to energy efficient vehicles (particularly automobiles), (d) adoption of energy saving measures in new and old buildings (e.g. well-insulated roofs, walls, windows), (e) transition to energy-efficient appliances (such as washing machines, dishwashers, refrigerators), (f) use of energy-efficient office equipment (such as computers and printers), (g) efficient land use systems that discourage biomass depletion and encourage carbon capture (sequestration), and many other measures. Individual citizens may play an important role in reducing greenhouse gas emissions by changing their energy consumption patterns. Some of the measures listed above would require government interventions at county, state and federal levels through administrative and legislative initiatives.

Emission reductions

The 1997 Kyoto Protocol was the first major international treaty to cut down greenhouse gas emissions. Although President Clinton (1992-2000) emphasized the need for climate change actions, he "never even brought the Protocol up for a vote" (*The Wall Street Journal*, Eastern Edition, 21 June 2005). The Republicans, in general, have opposed legislations requiring industries to cut down carbon dioxide emissions. Thus, when George W. Bush was elected as the President in 2000, the Republicans found a powerful ally on their side. The

Bush Administration rejected the basic tenet of the Kyoto Protocol requiring "industrialized countries to reduce heat-trapping smokestack and tailpipe emissions" (*The New York Times*, 5 August 2003). The smokestack emissions from coal-fired plants can be traced back to the industrial revolution some 200 years ago. Although oil (petroleum) now accounts for more than double the amount of energy produced by coal, the latter continues to be a major source of industrial pollution because of its abundance and its relatively low coast. With a proven reserve of 261 billion tons, the United States has the largest coal reserve in the world. It ranks second only to China in its annual production (1.1 billion tons per year versus 3 billion tons per year by China) (IER 2014). Limiting greenhouse gas emissions would require limiting both "smokestack and tailpipe releases of carbon dioxide [automobile exhausts], the main heat-trapping greenhouse gas". (*The New York Times*, 5 August 2003).

Although Bush tinkered with several climate change initiatives none of them was consequential. In 2003, a significant bipartisan initiative, McCain-Lieberman Climate Stewardship Act failed a vote in the US Senate (www.c2es.org/federal/congress/108/summary-mccain-lieberman-climate-stewardship-act . . . retrieved on 5 September 2016). To be fair on Bush and the Republicans, none of the three Presidents of the past two decades (i.e. Clinton, Bush and Obama) presented the Kyoto Protocol to the Congress for a formal ratification. Such resistance to the Kyoto Protocol has been "preventing the reduction of carbon emissions" (*The Guardian* [UK], 12 December, 2002). Against this backdrop, many states have been taking pragmatic initiatives to reduce greenhouse gas emissions, often irrespective of the party affiliations of political leaders championing these measures. Newspaper reports are replete with examples of such measures. For example, the *USA Today* (19 November, 2002) cited several such measures:

- California demanded "sharp reductions in carbon dioxide and other greenhouse gases emitted by vehicles".
- "Massachusetts ordered its older plants to curb harmful pollutants".
- "Nebraska ordered farmers to change planting practices so crops and soil retained more carbon, a way to reduce carbon-dioxide emissions that would help trap heat in the Earth's atmosphere.

- To confront global warming-induced sea level rise, New Jersey adopted a plan to curb greenhouse-gas emissions from in-state sources.
- "Traditionally laissez-faire New Hampshire . . . ordered its power plants to roll back carbon dioxide-emissions to 1990 levels because of evidence of climate change that could threaten the maple-syrup industry, skiing and trout fishing".

Some states were "even copying each other's good ideas". Illinois, North Dakota, Oklahoma and Wyoming had enacted farm laws similar to Nebraska's. Other windswept states were eying Texas' example in wind power (*USA Today*, 19 November 2002). Governors of both parties have been taking the lead on finding ways to reduce greenhouse gas emissions. However, some of the governors who were willing to promote climate change initiatives had been frustrated by the lack of progress at the federal level. Former California governor, Arnold Schwarzenegger (Republican) was upset that, as of May 2007, the federal government had not approved his state's "plan to limit tailpipe emissions of carbon dioxide from cars, light trucks, and sport-utility vehicles. Under the federal Clean Air Act, California may enact its own air-pollution standards as long as the US Environmental Agency (EPA) grants the state a waiver. Other states may then adopt California's tougher standards" (*The Christian Science Monitor*, 24 May 2007). In 2007, the nonpartisan National Governors Association detailed efforts made at the state level to reduce greenhouse-emissions through conservation as well as developing more renewable energy sources (*The Christian Science Monitor*, 24 May 2007).

Election of President Obama in 2008 was a major watershed in federal government's climate change initiatives, though following the loss of Democratic majority in the Congress in 2010 most of his efforts have been blocked by the Republicans. In the absence of legislative support by the Congress the President had only limited options to undertake climate change initiatives. Within the first 100 days of taking office in 2009, for instance, the President could have signed "an executive order that sets zero net-emissions goals for federal buildings. He could order the Environmental Protection Agency to begin regulating greenhouse gases. Or he could require federal agencies to include greenhouse-gas emissions among the effects they must report when weighing the environmental impact of their projects" (*The Christian Science Monitor*, 13 August 2008). In 2009, the House of Representatives (still controlled by the Democrats) passed the American Clean Energy and Security Act, but subsequently it died in the Senate. The Act would require "greenhouse gas emission to drop 17% by 2020 from 2005 levels. Companies that produced emissions, such as oil refineries, factories and electric utilities, would be required to either cut their emissions or buy pollution permits" (*USA Today*, 3 September 2009). The central theme of this Act was cap-and-trade, also called emission trading, which is a policy-based mechanism for controlling greenhouse gas emissions. Under this mechanism, "a government sets a cap—a limit—on the amount of greenhouse gas emissions various industries can emit into the atmosphere. This limit is gradually reduced over time to decrease total pollution levels" (*The Star*: https://www.thestar.com/news/ Canada/2015/04/13/what-is-cap-and-trade.html/ . . . retrieved on 21 August 2016). The cap and trade system is alternately called emission trading because it allows industries that succeed in cutting their emission levels below the set cap (i.e., allowable maximum emission levels) to sell their unused emission amounts to other industries that might not be able to meet their caps. Besides President Obama, Hillary Clinton, the 2016 presidential candidate, supports "cap and trade, under which companies can trade carbon credits to reduce harmful emissions" (*The New York Times*, 13 April 2015). One of the flexibilities of the system is that it can be applied within a state or between several states or internationally between two or several countries or even between a US state and the administrative unit of another country (such as the one being proposed between California and the Canadian provinces of Ontario and Quebec). At the international level, "emission trading allows industrialized countries to purchase 'assigned amount units' of emissions from other industrialized countries that find it easier, relatively speaking, to meet their emission targets" (Houghton 2004, 248).

Perhaps the most basic philosophical issues of climate change have been articulated by President Obama in his second term (on 25 June 2013): "Someday, our children, and our children's children, will look at us in the eye and they'll ask us, did we do all that we could when we had the chance to deal with this problem and leave them a cleaner, safer, more stable world?" (https://www.whitehouse.gov/energy/climate-change . . .

retrieved on 2 September 2016). In June 2013, President Obama outlined the Climate Action Plan—"the steps his Administration would take to cut carbon pollution, help prepare the United States for the impacts of climate change, and continue to lead international efforts to address global climate change" (https://www.whitehouse.gov/energy/climate-change . . . retrieved on 2 September 2016). Reductions of carbon dioxide emissions constitute perhaps the central issues of this action plan. Among several measures to cut down carbon dioxide emissions it has put in place the first-ever carbon pollution standards for power plants, which account for roughly one-third of all domestic greenhouse gas emissions (https://www.whitehouse.gov/energy/climate-change . . . retrieved on 2 September 2016).

In public discourse on greenhouse gas emission reductions coal-fired plants have received significant attention from politicians. For example, in 2015 President Obama unveiled "a final set of regulations to cut planet-warming carbon pollution from coal-fired plants" (*The New York Times*, 23 April 2015). The Republicans called the rules "a war on coal that could lead to shutdown of hundreds of coal plants" (*The New York Times*, 23 April 2015). The U.S. Supreme Court agreed with the Republicans in a majority decision delivering a major blow to Obama's plan by "putting on hold federal regulations to curb carbon dioxide emissions from coal-fired plants, the centerpiece of his administration's strategy to combat climate change" (Hurley and Volcovici 2016).

Energy efficiency

Energy efficiency is "using less energy to provide the same service" (Lawrence Berkeley National Laboratory: eetd.lbl.gov/ee/ee-1.html . . . retrieved on 18 August 2016). Familiar examples include energy-efficient homes, high-efficiency furnaces, refrigerators, washing machines (for clothes), computers and printers, to name a few. By reducing energy consumption, high-efficiency appliances reduce energy costs (energy bills) and reduce the amounts of greenhouse gas emissions. Appliance energy regulations constitute a major policy measure for controlling or minimizing household and commercial greenhouse gas emissions.

The Building Technologies Office (BTO) "implements minimum energy conservation standards for more than 60 categories of appliances and equipment" (energy.gov/eere/buildings/

appliances-and-equipment-standards-program . . . retrieved on 21 August 2016). The Appliance and Equipment Standards Program "issues regulations for appliance and equipment standards and test procedures, and for implementation, certification, and enforcement" (energy.gov/ eere/buildings/appliances-and-equipment-standards-program . . . retrieved on 21 August 2016). This program, administered by the U.S. Department of Energy (DOE), covers more than 60 products, representing about 90% of home energy use, 60% of commercial building energy use, and 30% of industrial energy use (energy.gov/eere/ buildings/appliances-and-equipment-standards-program . . . retrieved on 21 August 2016).

Confronted with the Republican opposition to "regulatory overkill", the Obama administration has a mixed record on appliance energy regulations. "With a sweeping climate bill having died in the Senate in Mr Obama's first term, his only options for major action on the issue in the second term appear to involve executive action" (*The New York Times*, 13 June 2013). Yet, the "Whitehouse has blocked several Department of Energy regulations that would require appliances, lighting and buildings to use less energy and create less global warming pollution, as part of a broader slowdown of new antipollution rules issued by the Obama administration" (*The New York Times*, 13 June 2013). Regulatory review times at the Whitehouse Office of Management and Budget (OMB) "are now the longest in 20 years, having spiked sharply since 2011" (*The New York Times*, 13 June 2013). Some of the Congressional Democrats and environmental advocates have criticized these delays, alleging that "Mr Obama has not made good on his recent promises and has failed to show urgency about climate change" (*The New York Times*, 13 June 2013). As implied at the beginning of this paragraph, politics seem to be an underlying factor for such delays. The Whitehouse, "sensitive to Republican charges that it was threatening the economy by pushing out dozens of so-called job-killing regulations, reined in the [review] process last year, leaving many rules awaiting action for months beyond legal deadlines (*The New York Times*, 13 June 2013). One example of a delayed rule is a DOE "proposal to mandate greater energy efficiency for walk-in coolers and freezers, which are common at many restaurants, supermarkets and warehouses . . . The delay has added more than 10 million metric tons of carbon dioxide to the atmosphere, according to calculations by two independent watchdog groups" (*The New York*

Times, 13 June 2013). By slowing down the regulatory review process, the Obama administration sought "to maintain a balance between our obligation to protect the health, welfare and safety of Americans and our commitment to promoting economic growth, job creation, competitiveness and innovation" (*The New York Times*, 13 June 2013).

While the federal government is bogged down with the implementation issues of its own appliance energy regulations, at local levels—far from Washington, DC—some of the civic leaders have been apolitical in handling greenhouse gas emission issues. For example, James Brainward, a Republican mayor of Carmel, Indiana, has sought to be proactive on climate change issues. "The city has reduced its energy use with fuel-efficient city cars and small trucks, LED lighting and so-called green buildings. It also pipes the methane gas from the treatment of wastewater into boilers that help produce so-called biosolids that can be used as fertilizer" (*The New York Times*, 25 October 2014).

Natural gas-fired plants

Natural gas is an anthropogenic greenhouse gas contributing to climate change. However, compared to coal or oil (petroleum) it is a much cleaner fuel. Burning natural gas produces "25 to 40 per cent lower [amounts of greenhouse gas emissions] per unit of generated electricity." (Suzuki Foundation: http://www.davidsuzuki.org/issues/climate-change/science/energy/natural-gas/ . . . retrieved on 25 August 2016). From emission mitigation perspectives natural gas may be considered at best as a "transition fuel" because of the high cost of pipelines required for its transport from the source to the plants. As demand continues to soar for natural gas-fired power plants large numbers of such plants may be counter-productive in the sense that their cumulative emissions may cancel out the benefit of per capita reduction of emissions.

In the United States the proponents of climate change mitigation often cite natural gas-fired power plants as a short-term solution for emission reductions. Even many Republicans, especially some of the new generation GOP leaders, support natural gas-fired power plants. George P. Bush (son of Jeb Bush), Texas Land Commissioner, is an example. In a stunning departure from the official GOP positions on climate change he has turned out to be a staunch critic of coal-fired power plants. He has also defended the EPA, contrary to some of the scathing criticisms of this federal regulatory agency by several GOP leaders. "Regardless of your politics," Mr Bush said, "the EPA is regulating coal and ratcheting down its overall usage in our electricity grid." . . . Because of that and concerns about global warming, Mr Bush said, "Texas should move away from coal-fired power and transition to a natural gas-based energy economy and then, in the long term, renewables." (*The New York Times*, 31 August 2914). Such views are very similar to the public statements of his father (Jeb Bush) "on the advantages of natural gas from a carbon emissions standpoint . . . His grandfather (George H.W. Bush) was an early actor on pushing for global climate treaties more than two decades ago." (*The New York Times*, 31 August 2914). Kevin McCarthy (GOP), the current House Majority Leader (2014-) from California is also on record stating that "climate pacts aren't the best use of our money" and arguing that "transitioning to natural gas is a better way to reduce greenhouse gas emissions" (*The Christian Science Monitor*, 30 November 2015).

Adaptation to climate change

Adaptation to climate change refers to strategies for coping with actual or expected impacts of climate change. More formally, the IPCC defines it as "adjustment in natural or human ecosystems in response to actual or expected climatic stimuli or their effects, which moderates harm or exploits beneficial opportunities" (http://www.vcccar.org.au/climate-change-adaptation-definitions . . . retrieved on 26 August 2016).

Adaptation to sea level rise

Some of the adaptation measures are already in place in different regions of the United States, mostly as parts of flood control measures and protection of low-lying coastal areas from rising sea levels. The latter seems to be the most popular climate change adaptation measures. For example, in 2014 a bipartisan group of political leaders in Virginia considered "ways to adapt the low-lying Hampton Roads region in Virginia to science-based predictions that the seas will rise at least a foot in 30 years and five feet or more by the end of the century . . . At the heart of the Hampton Roads area is Norfolk, home to the world's largest naval base and a city that the White House warned in May [2014] is among the nation's most vulnerable to rising sea" (*The New York Times*, 7 July 2014). In the Florida Keys, possible responses to sea level rise include elevating roads and switching the Bermuda grass at the local golf course to salt-tolerant varieties of *paspalum* grasses (*The New York Times*, 25 October 2014).

In the wake of Hurricane Sandy that generated extensive storm surge flooding in New York, city and state officials "are beginning to consider longer-term solutions to prevent a recurrence of the flooding" (*The Christian Science Monitor*, 13 November 2012). Among preventive measures planners and engineers have considered the idea of stopping the storm surge by a five-mile long barrier outside the harbor. In 2009, three years prior to Sandy, a British company (Halcrow Group) that works on infrastructure projects worldwide proposed such a structure: "a five-mile fixed barrier stretching from Sandy Hook, NJ to Breezy Point in the Big Apple Borough of Queens" (*The Christian Science Monitor*, 13 November 2012). One proponent of some form of sea wall is Malcolm Bowman of the State University of New York at Stony Brook, who heads the school's group that predicts and models storm surges for the New York area (*The Christian Science Monitor*, 13 November 2012). Other more innovative ideas have also been considered. For example, "what might have happened if New York had built marshes and oyster beds at the tip of Manhattan" that could absorb some of the storm surge energy? (*The Christian Science Monitor*, 13 November 2012).

In 2014, "Massachusetts state officials formally announced a $50 million effort to adapt to current and future effects of global warming, with some 80 percent of the money slated to support projects that will reduce the vulnerability of critical facilities in cities and towns to power failure" (*The Christian Science Monitor*, 14 January 2014). An additional amount of $10 million was allocated for infrastructure projects, such as rebuilding sea walls and repair or remove dams to deal with coastal and inland flooding from severe storms. "The program typifies a shift that has been under way during the past several years as states with global-warming policies have added adaptation to their agendas . . . More recently, they started to move beyond studies of needs to initiating projects . . . So far, 15 states—mainly along the East and West Coasts—have completed adaptation plans . . . Another four are still developing plans, while seven states and the District of Columbia have yet to follow through on recommendations to develop adaptation plans contained in broader climate action documents." (*The Christian Science Monitor*, 14 January 2014). In Alaska, Obama administration has supported several climate change adaptation projects, including relocation of coastal villages that are vulnerable to impacts of rising sea levels. In 2015, Mr Obama announced that "the Denali Commission, the federal agency that coordinates government assistance to communities in Alaska, would oversee short- and long-term programs to safeguard and repair the coastal villages, including 'voluntary relocation efforts, where appropriate' . . . The assistance package will include new grant programs from the Agriculture Department and the Environmental Protection Agency for water and waste projects in vulnerable Alaskan villages" (*The New York Times*, 3 September 2015).

Despite good progress in adaptation plans by many states, there are significant differences in the way each state interprets such plans. For example, "in Massachusetts, California, or New York, the link between adaptation projects and climate change is explicit. In other states, where politicians give global warming a cold shoulder, the projects that could fit into an adaptation bin are merely labeled coastal or wetlands restoration projects or projects to build a more resilient electrical grid. The impact on resilience can be the same, but the inconsistent labeling can cloud attempts to keep track of dedicated adaptation efforts" (*The Christian Science Monitor*, 14 January 2014). To deal with climate change impacts effectively, "countries will need to adopt what Stanford University's Christopher Field refers to as 'a portfolio of measures' for adaptation to change, as well as to reduce greenhouse-gas emissions and land-use practices that have compounded the change" (*The Christian Science Monitor*, 11 March 2010). Across the United States, a growing number of state and local governments are pulling together to deal with the effects of climate change, "as a new tracking tool from the Georgetown Climate Center at Georgetown University Law Center shows . . . The Obama administration, hoping to build on momentum at the local level, has created a task force of state and local officials who are active on the issues" (*The New York Times*, 25 October 2014).

Renewable energy

In addition to the preceding adaptation projects, security of "energy supplies and delivery is a cornerstone of the national adaptation strategy President Obama unveiled last year [in 2013]" (*The Christian Science Monitor*, 14 January 2014). This helps explain why the large portion of Massachusetts' electric gridline projects "aimed at boosting the resilience of the grid and introducing micro-grids built on clean energy technologies" (*The Christian Science Monitor*, 14 January 2014). Similarly, in Alaska, the US Department of Energy announced "initiatives to help remote Alaskan communities and native tribes develop clean-energy programs to reduce their reliance on fossil fuels" (*The New York Times*, 3 September 2015). In Texas, the state has been pushing for more than decade for the development of renewable energy sources, such as wind and solar energy. As a result, the wind power had "jumped six-fold in three years" by 2002 (*USA Today*, 19 November 2002). From the very beginning of his administration, Mr Obama and his first-term Energy Secretary, Nobel Prize winner Stephen Chu have been championing climate change adaptation through the development of renewable energy sources. The program also included "advanced biofuels and clean coal, which is especially processed to reduce carbon emissions" (*USA Today*, 2 March 2009). Wind and solar power have tremendous potentials for future growth. Wind farm turbines, such as those in Colorado, Texas, Iowa, South Dakota, are the nation's fastest-growing renewable energy source. In 2003, it provided only 1% of US power but it increased by 177% from 2003 to 2007 (*USA Today*, 2 March 2009). In 2013, about 4% of the electricity generated in the United States was produced from wind energy (http://windenergyfoundation.org/about-wind-energy/faqs . . . retrieved on 7 September 2016). Solar power is perhaps the most promising long-term renewable energy source. "Nanotechnology advances promise better collection of the sun's energy" (*USA Today*, 2 March 2009). Nuclear power may not be a renewable energy source but it does not contribute to greenhouse-gas emissions. Its production has increased from about 8% of the total US electricity sources in the early 1970s to about 20% in 2015 (http://www.nei.org/Knowledge-Center/Nuclear-Statistics/US-Nuclear-Power-Plants . . . retrieved on 7 September 2016). Despite such rapid expansion of nuclear power plants throughout the United States their further expansion is constrained by the high cost of construction of the plants and problems with safe disposal of nuclear waste materials (*USA Today*, 2 March 2009).

To meet anticipated longer-term growth in energy demand, several governors are leading efforts to "conserve energy sources while actively seeking to diversify supplies by expanding renewable sources, including energy generated from solar, wind, hydropower, geothermal, and biomass. In addition to providing protection against price volatility, these efforts can also reduce greenhouse-gas emissions" (*The Christian Science Monitor*, 24 May 2007). Former California governor Arnold Schwarzenegger is one of the vocal proponents of transition to renewable energy sources. In 2015 United Nations Climate Change Conference at Paris (COP 21), he argued that fossil fuels would eventually run out and therefore it was "wise to start investing in and transitioning to renewable energy sources" . . . As governor of California, Schwarzenegger oversaw California's transitioning to sourcing 33 percent of its energy from renewable like solar power by 2020, and the passage of bills aimed at reducing California's greenhouse-gas emissions" (*The Christian Science Monitor*, 11 December 2015). In Paris, Schwarzenegger also attracted worldwide attention for endorsing "eating less meat as an additional step towards helping the planet breathe a bit easier. Meat producers like cows are a major source of greenhouse-gas emissions" (*The Christian Science Monitor*, 11 December 2015). Renewable energy has been gaining attention even at local levels in some of the energy-rich states. For example, Georgetown, an oil-rich town in Texas, an unlikely place to ditch fossil fuels, is scheduled to get all its electricity by 2017 from renewable sources, such as wind and solar power. Dale Ross, mayor of Georgetown asserted that "Make no mistake, this was a business case . . . Though nearly everybody has an oil derrick in their backyard . . . sun and wind power are abundant and cheap, and don't experience the same price volatility as fossil fuels" (*The Christian Science Monitor*, 25 April 2015).

Climate change politics and policies

As data in Table 8.7 indicate, approximately one-half of the analyzed text dealt directly with climate change politics and policies. Based on the contents of the selected newspaper reports, we have identified some of the major themes of political news on climate change.

Federal government's ambivalent positions on climate change

At the United Nations Framework Convention on Climate Change (UNFCCC) signed by over 160 countries in Rio de Janeiro in June 1992 the world agreed "to prevent dangerous climate change. Countries have since worked on mandatory emissions curbs in what is known as the Kyoto Protocol" [signed in December 1997] (*USA Today*, 10 February 2006; Houghton 2004). The Kyoto Protocol was the first international treaty "that required industrialized countries to reduce heat-trapping smokestack and tailpipe emissions" (*The New York Times*, 5 August 2003). Despite its theoretical support for this landmark treaty, from the very beginning the US federal government took an ambivalent position on this initiative largely because of deep-seated concerns about potentially harmful effects of the treaty on the US economy. In particular, the Republicans and powerful business lobbies opposed the treaty. At the pre-Kyoto Buenos Aires negotiations (in November 1997), the United States signed the Kyoto Protocol. Contradicting this historic commitment to reduce emissions of six greenhouse gases up to 7% below what they were in 1990 by 2008, the Clinton administration never intended to submit the treaty to the Senate to be ratified (http://www. agiweb.org/legis105/climate.html#admin . . . retrieved on 14 September 2016). Several Republican members of the Congressional delegation to Buenos Aires criticized such ambivalent move: "In effect, the Clinton Administration is trying to have both ways. The interpretation by other nations will be that the US intends to implement the Kyoto Protocol. But by not submitting the treaty immediately to the Senate for ratification, the Administration is attempting to tell the American people . . . 'Don't worry. Signing this document doesn't mean anything'" (American Geological Institute Global Climate Change Update [12-29-98], http://www.agiweb.org/legis105/climate.html#admin . . . retrieved on 14 September 2016). Besides the Kyoto Protocol, the Republicans had opposed several other greenhouse-gas emission reduction initiatives proposed by President Clinton. When President Bush took over the federal government in 2001 not only he inherited this uncertainty in climate change initiatives he doubled down on this uncertainty by opposing the Kyoto Protocol. By that time, powerful industry and business groups, such as the Global Climate Coalition (GCC) and the International Climate Change Partnership (ICCP) "showered more than $300 million on both parties since 1990 to block federal action, according to federal records compiled by the non-partisan Center for Responsive Politics" (*USA Today*, 19 November 2002).

State governments' leadership in climate change

In short, the Clinton administration failed to initiate any significant climate change initiatives. Because of President Bush's stated opposition to the Kyoto Protocol the leadership vacuum in climate change at the federal level deepened further. Countering federal lack of leadership, this vacuum has partly been filled up by states. Somewhat paradoxically, Texas is a good example. When George Bush was the Governor of Texas his state was "among an increasing number of states that are tackling climate change problems in the absence of action from Washington. Their initiatives represent sensible measures that don't cause economic harm or impose harsh penalties on industry—concerns the Bush administration raises to justify doing nothing" (*USA Today*, 19 November 2002). A report from the Pew Center on Global Climate Change, a research group with business backing, documented initiatives during the 1990s "even in states not known for environmental zeal" (*USA Today*, 19 November 2002). However, many powerful industries that had earlier blocked legislations in Washington were "fighting a rear-guard action in the states. Auto companies for example, [were] trying to kill California's drive for tighter standards, while the coal industry in West Virginia persuaded the legislature to bar greenhouse-gas initiatives. More often, Republicans and Democrats have united in state houses to overcome business resistance and show how efforts to combat global warming can be good for the environment and the economy. If only Washington would catch on" (*USA Today*, 19 November 2002).

Unquestionably, California has been at the forefront of fighting climate change. In particular, it has been a leader in motor vehicle pollution control. The latter has been a bipartisan initiative, which has been championed by Arnold Schwarzenegger (Republican Governor of California: 2003-2010) and Jerry Brown (Democratic Governor of California: 2011-current, 1975-1983, also California Attorney General: 2007-2011). In 2006, the passage of AB 32, the California Global Warming Solutions Act of 2006, a signature legislative initiative by Schwarzenegger "marked a watershed moment in California's history. By requiring in law a sharp reduction of greenhouse gas (GHG) emissions, California set the stage for its transition to a sustainable, low-carbon future" (https://www.arb.ca.gov/cc/ab32/ab32.htm . . . retrieved on 30 September 2016). From the very beginning, California has been at odds with the federal government which has been stonewalling this landmark initiative. At an EPA hearing (on 22 May 2007) to consider the request by California and ten other states, California Attorney General Jerry Brown threatened to sue the EPA for blocking California's vehicle pollution plan (*The Christian Science Monitor*, 24 May 2007). In a public radio broadcast in May 2007 (from KPBS station, San Diego) Mr Brown commented: "In ordinary politics they are talking about immigration and Iraq—very serious—but from a long-term perspective, the threat of oil dependency and global climate disruption is much more threatening, much more difficult to deal with, and we gotta get going" (*The Christian Science Monitor*, 24 May 2007). As expected, the auto industry has been fighting this legislation. A lone voice of opposition at the EPA hearing came from an auto industry representative, Steve Douglas of the Alliance of Automobile Manufacturers. In an Associated Press (AP) account of the hearing, Mr Douglas complained that "a patchwork of state-level fuel economy regulations, as is now proposed by California, is not simply unnecessary, it's patently counterproductive . . . The state's waiver request contains many assumptions and undocumented claims about its benefits in countering global warming" (*The Christian Science Monitor*, 24 May 2007).

Writing in *The Washington Post* (21 May 2007), Arnold Schwarzenegger and Jodi Rell (Republican Governor of Connecticut: 2004-2011) blasted the EPA and the Bush Administration for "inaction and denial" on climate change, which they said, "borders on malfeasance" (*The Christian Science Monitor*, 24 May 2007). "By continuing to stonewall California's request, the federal government is blocking the will of tens of millions of people in California, Connecticut, and other states who want their government to take real action on global warming" (*The Christian Science Monitor*, 24 May 2007). Jon Huntsman, Jr, the Republican Governor of Utah (2005-2009), announced in may 2007 that his "state would join five others (California, Arizona, New Mexico, Oregon, and Washington) and the Canadian province of British Columbia as part of the recently formed Western Regional Climate Action Initiative . . . This isn't about party politics, Governor

Huntsman said in a *Salt Lake Tribune* story about the announcement. It's about doing the right thing for all of our citizens" (*The Christian Science Monitor*, 24 May 2007). Many cities are also forging ahead of Washington on motor vehicle pollution control initiatives. In New York, for example, Mayor Michael Bloomberg (2002-2013) announced that "every yellow cab in the city will be a fuel-efficient hybrid model by 2012 . . . Once the new standards are fully in place, carbon emissions will be reduced by more than 200,000 metric tons per year, city officials said" (*The Christian Science Monitor*, 24 May 2007).

Climate change deniers versus activists

The Republican-Democratic divide on climate change issues is further reinforced by two competing political forces. Perhaps, the most powerful force is a climate change denial movement which is supported by a number of well-funded conservative foundations with "so-called dark money, or concealed donations" (Fischer 2013). Most of these organizations are affiliated with and funded by powerful business lobbies representing coal, oil, and gas industries. The main purpose of these organizations is "to sow doubt about the facts of global warming . . . These organizations play a key role in the fossil fuel industry's 'disinformation playbook', a strategy designed to confuse the public about global warming and deny action on climate change. Why? Because the fossil fuel industry wants to sell more coal, oil, and gas— even though the science clearly shows that the resulting carbon emissions threaten our planet" (UCSUSA: Union of Concerned Scientists, USA 2016: http://www.ucsusa.org/global_warming/solutions/fight-misinformation/global-warming-skeptic.html#.V_bAZeArLIU/ . . . retrieved on 6 October 2016). According to an independent study, at least 140 foundations funneled $558 million to almost 100 climate change denial organizations from 2003 to 2010 (Fischer 2013). The following is a list of some of the high-profile climate change denial organizations (listed in alphabetic order in the website of the Union of Concerned Scientists (http://www.ucsusa.org/ . . . retrieved on 6 October 2016).

- American Enterprise Institute (AEI)
- Americans for Prosperity (AFP)

- American Legislative Exchange Council (ALEC)
- Beacon Hill Institute (BHI) of Suffolk University
- Cato Institute
- Competitive Enterprise Institute (CEI)
- Heartland Institute
- Heritage Foundation
- Institute for Energy Research (IER)
- Manhattan Institute for Policy Research

Some of these organizations acknowledge that global warming is real but question human role in warming. In other words, they challenge the anthropogenic global warming hypothesis. Others reject the global warming hypothesis altogether, asserting that "the very question of whether the climate is warming is in doubt" (http://www.ucsusa.org/ . . . retrieved on 6 October 2016). Kansas oil and gas billionaires Charles and David Koch (Koch brothers), who are co-founders of the Cato Institute and Americans for Prosperity believe that "climate change is a left-wing hoax". They have spent at least $88.8 million directly supporting 80 climate change denial groups since 1997 (www.greenpeace.org/ . . . retrieved on 4 October 2015; *The New York Times*, 21 September 2010). "Of that total, $21 million went to groups that recently bought a full page *New York Times* advertisement defending Exxon Mobil from government investigations into its systematic misrepresentation of climate science" (http://www.ecowatch.com/koch-brothers-continue-to-fund-climate-change-denial-machine-spend-21m-1891180152.html/ . . . retrieved on 4 October 2016). The Koch Industries conglomerate ranks as "one of the top 10 air polluters in the country" (*The New York Times Blogs*, 16 November 2015). In 2006, when California passed its automobile emission control law, known as AB 32, Koch brothers spent about $1 million unsuccessfully trying to kill the legislation. Two other Texas-based oil and gas companies, Valero and Tesoro, contributed a large chunk of $8.2 million raised to support an unsuccessful ballot proposition to defeat the legislation (*The New York Times*, 21 September 2010).

Compared to the climate change denial movement, which is supported largely by industries and business interests, climate change activists represent a grass-roots network of perhaps thousands of organizations of different sizes, which are funded partly by millions of individual donations and partly by environmentally-friendly philanthropic foundations. The main goal of these organizations is to pressure policymakers to address climate change issues, especially reductions of greenhouse gas emissions. The following are some of the leading climate change activist organizations:

- Greenpeace
- Sierra Club
- 350.org (co-founder: Bill McKibben)
- Climate Reality Project (founded by Al Gore)
- Union of Concerned Scientists (Yale Project on Climate Change Communication)
- Next Gen (founded by California billionaire Tom Steyer)

Some of these organizations have millions of supporters. Al Gore's Climate Reality Project has a network of 5 million individuals (http://www.climaterealityproject.org/ . . . retrieved on 6 October 2016). McKibben's 350.org similarly has millions of supporters in both USA and 188 other countries (https://350.org/ . . . retrieved on 11 October 2016). Although most of the climate change activist organizations rely heavily on relatively small donations from grassroots supporters some of them receive funding from high-profile philanthropic organizations. For example, Greenpeace, Sierra Club, and 350.org have posted in their websites samples of donations from several foundations that range from as small as $100 to as high as $1 million or more. Some of the high-profile donors include Rockefeller Brothers Fund, John D. & Catherine T. MacArthur Foundation, David L. Klein, Jr. Foundation, Blue Moon Fund, and others (www.greenpeace.org/ . . . retrieved on 4 October 2015). Overall, there are very few industrial or business interest backers. The only exception is the Next Gen, founded by Tom Steyer, the billionaire behind Democrats and the crusader against global warming. Among all climate change activists, the Next Gen has received most extensive media coverage, probably because its founder happens to be a billionaire industrialist and at the same time a Democrat whose climate change activist philosophy stands in sharp contrast to the competing climate change denial philosophy of industrialists Koch Brothers.

Tom Steyer is a retired hedge fund manager who had used some of his fortune in an effort to make climate change a potent political issue in 2014 Congressional elections in pivotal states like Florida, Iowa and Virginia (*The New York Times*, 18 may 2014). In 2013, Steyer "spent $11 million to help elect Terry McAuliffe governor of Virginia and millions more intervening in a Democratic congressional primary in Massachusetts." (*The New York Times*, 18 February 2014). His super-PAC also invested heavily to oust the former Republican Pennsylvania governor and elect the current Democratic governor in 2014. The Next Gen also helped in shaping several state legislative races in California, Washington and Oregon (*USA Today*, 15 October 2014).

During his campaign for the Democrats Mr Steyer said that "his multimillion-dollar crusade to slow global warming rests on exposing the human consequences of fossil-fuel consumption." . . . Further, one of his goals is "to disrupt American politics by making climate change a wedge issue in campaigns" (*USA Today*, 15 October 2014). In all, "Steyer has plowed more than $42 million of his fortune into federal campaign accounts since early March 2013, making the San Francisco Democrat the largest super-PAC donor of the 2014 election. His political organization has opened 40 offices, built a team of 800 employees and volunteers in its targeted states and made contact with more than 1.5 million voters" (*USA Today*, 15 October 2014). In early February 2014, "Mr Steyer gathered two dozen of the country's leading liberal donors and environmental philanthropists to his 1,800-acre ranch in Pescadero, California—which raises prime grass-fed beef—to ask them to join his efforts. People involved in the discussions say Mr Steyer is seeking to raise $50 million from other donors to match $50 million of his own . . . He is seeking to build a war chest that would make his political organization, Next Gen Climate Action, among the largest outside groups in the country, similar to scale to the conservative political network overseen by Charles and David Koch." (*The New York Times*, 18 February 2014).

Religion and climate change politics

The role of religion in U.S. politics has evolved from a long-held "Christian axiom that nature has no reason for existence save to serve man" to a more contemporary and less divisive efforts of some [Christian] leaders to frame "the problem of climate change as a matter of religious morality" (Kolmes and Butkus 2007). In 2010, the Pew Research Center conducted a survey of U.S. adults and found that "81% of all adults, including strong majorities of all major religious traditions, favored 'stronger laws and regulations to protect the environment,' while 14% opposed them" (Pew Research Center 2016: http://www.pewinternet.org/2015/10/22/religion-and-views-on-climate-and-energy-issues/ . . . retrieved on 12 October 2016). Attitudes of different religious groups have been consistent with this survey. For example, in 2007, when the IPCC Fourth Report (AR 4) indicated that carbon dioxide emissions had been spiraling upward faster than earlier predicted, leaders of religious groups, among others, urged President Bush and Congress to take action against global warming (*The Christian Science Monitor*, 24 May 2007). "In an open letter May 22 [2007], as reported by Reuters and other news sources, more than 20 Christian, Jewish, and Muslim groups declared that climate change is a 'moral and spiritual issue'. The religious leaders added: 'Global warming is real, it is human-induced, and we have responsibility to act.'" (*The Christian Science Monitor*, 24 May 2007). A subsequent interview-based survey in fall 2007, conducted by Baylor University among 1700 adults, found that most respondents agreed that "if we do not change things dramatically, global climate change will be a disaster (67%); coal, oil and natural gas will be exhausted (70%), and most plant and animal life will be destroyed (57%)" (*USA Today*, 18 September 2008). Another significant finding of this study was that, contrary to the "myth of the evangelical environmental movement", a majority of the evangelical Protestants (55%) were less likely to be alarmed about global climate change projections, compared to 49% of other religious groups. Similarly, 41% of evangelicals indicated that the government was already spending too much on the environment, whereas 81% of the Jews and 79% of other religious groups complained that the government spending for the environment was too little (*USA Today*, 18 September 2008). The survey coordinator, Baylor University sociologist F. Carson Mencken, reflected on this finding: "This is not to say that evangelicals are anti-environment but their support for environmental issues is not as strong as other religious traditions." (*USA Today*, 18 September 2008). Indeed, environmentalism has been controversial among evangelicals. "When the National Association of Evangelicals launched a 'Call to Action' on climate change in 2006, some religious conservatives, led by James Dobson of Focus on the Family, strongly opposed it" (*USA Today*, 18 September 2008).

Perhaps, the most significant position on the role of religion in climate change politics has been articulated by Pope Francis (in June 2015) when he issued a highly anticipated and highly controversial *encyclical* declaring that climate change is "a moral imperative" (*The New York Times*, 17 June 2015; *The Christian Science Monitor*, 18 June 2015). An encyclical is "a papal letter addressed to bishops and all members of the Roman Catholic Church" (Barber 1998, 460). It is a teaching document which is "among the strongest and most authoritative statements made by the Catholic Church" (*The New York Times*, 17 June 2015). Invoking impending disasters of climate change, the184-page encyclical, entitled "*Laudito Si*" (meaning "Praise be to You"), emphasized the moral imperative to "care for our common home" (i.e. the earth/environment), but to articulate his argument the document included several politically-charged messages on anthropogenic global warming and the resulting climate changes. These included (paraphrasing his messages reported in *The New York Times*, 17 June 2015):

- Climate change is an established science.
- Burning fossil fuels are warming the planet.
- Climate change impacts threaten the world's poor.
- Government policies should cut fossil fuel use.

Further, the encyclical sounded a note of moral urgency: "If present trends continue, this century may well witness extraordinary climate change and an unprecedented destruction of ecosystems, with serious consequences for all of us" (*The Christian Science Monitor*, 18 June 2015). To prove his point, he also described a "technological paradigm" in which industrialized countries recklessly pursue profits at the expense of the environment and the globe's poorest peoples. As a remedy to this situation, he called for a "bold cultural revolution" that would "recover the values and the great goals swept away by our unrestrained delusions of grandeur" (*The Christian Science Monitor*, 18 June 2015).

Many Roman Catholic bishops throughout the United States repeated those messages in their sermons hoping that the pope's encyclical on climate change would resonate with the church members, particularly with Roman Catholic politicians. For example, Thomas G. Wenski, the Roman Catholic Archbishop of Miami, planned "a summer of sermons, homilies and press events designed to highlight the threat that a warming planet, rising sea levels and more extreme storms pose to his community's poorest and most vulnerable" (*The New York Times*, 17 June 2015). Besides Florida, the Catholic bishops pushed pope's climate change message in other swing states for presidential election. In Iowa, the bishops of Des Moines and Davenport planned "a news media event at a wind turbine manufacturing facility", where they highlighted findings that "climate change drives the drought and floods that plague Iowa farmers" (*The New York Times*, 17 June 2015). Similarly, the bishops of Cincinnati and of Las Cruces, New Mexico, planned news conferences and events on climate change. The bishop of Sacramento, California, in a state in the grip of a record drought, planned an event highlighting the link between drought and climate change (*The New York Times*, 17 June 2015).

Most of the above activities and pope's statements seem to be consistent with mainstream discourse on the science of climate change, but the concept of anthropogenic climate change is by no means an uncontested hypothesis. Thus, perhaps unwittingly, Pope Francis ventured into the arena of climate change politics, especially in the U.S. politics. Several prominent Roman Catholic Republican presidential primary candidates, notably Jeb Bush and Marco Rubio, found themselves at odds with the Pope's position. Reacting to the Pope's encyclical on climate change, Jeb Bush, a devout Roman Catholic, told his campaign rally in New Hampshire: "I hope I'm not going to get castigated for saying this by my priest back home, but I don't get economic policy from my bishops or my cardinals or my pope" (*The Christian Science Monitor*, 18 June 2015). Other Roman Catholic Republican presidential candidates, including Marco Rubio, Rick Santorum and Bobby Jindal, expressed skepticism about human-induced climate change. Rubio, in particular, said "he believes that the Earth's climate is constantly changing but that humans are not responsible for climate change in the way some of these people out there are trying to make us believe" (*The New York Times*, 17 June 2015). However, the pressure to respond to the pope's position on climate change could be particularly intense for Mr Bush and Mr Rubio, both from Florida, which is highly vulnerability to climate change, especially to sea level rise. The findings of a 2014

National Climate Assessment, a scientific study by 13 federal agencies, identified Miami as one of the United States cities most vulnerable to physical and economic damage as a direct result of human-induced climate change (*The New York Times*, 17 June 2015).

Not only Pope's position on climate change poses a political challenge for the Republicans, "many wonder if the pope's efforts can do much to change the politics of climate change" (*The Christian Science Monitor*, 18 June 2015). "Most people, and most Catholics, view their politics through right or left secular lenses, not necessarily through what the pope has to say about anything," says Charles Camosy, professor of theology and an expert on Christian bioethics at Fordham University in New York (*The Christian Science Monitor*, 18 June 2015). A Pew Research Center survey, conducted in June 2015 (overlapping the period of the encyclical), found sharp differences of views along party lines regarding both the extent of global warming and its underlying causes. Whereas, 80 percent of the Catholic Democrats said that there was solid evidence that the Earth was warming, only about one-half of the Catholic Republicans held such a view. Perhaps more importantly, 60 percent of the Catholic Democrats believed that global warming was man-made and that it posed a serious problem, but only about one-quarter of Catholic Republicans agreed with these statements (*The New York Times*, 17 June 2015).

Like Bush and Rubio, "many conservative Catholics, already suspicious of the pope's emphasis on traditionally liberal social justice issues, have . . . begun to dismiss the church's moral teachings on the environment." (*The Christian Science Monitor*, 18 June 2015). "Though climate change is a very hyper-partisan issue, I'm not sure it's always framed as a moral issue as much as an economic debate about how to solve these problems, whether through regulation or private enterprise solutions," said Christopher Vogt, chair of the department of theology and religious studies at St. John's University in New York. He added: "I am hopeful that raising it as a moral issue, and a spiritual issue, it might bring additional people into the debate who otherwise had not been interested or engaged in climate change or environmental issues." (*The Christian Science Monitor*, 18 June 2015). Besides, some of the Catholic theologians insist that Catholic teachings about the environment and climate change are not necessarily entirely new issues. These go back to Pope Francis's predecessors John Paul II and Benedict XVI, who was known as the "green pope" for his environmental efforts. Still, "*Laudito Si* is the first papal encyclical devoted entirely to the moral imperatives of caring for the environment" and the timing of the document seemed to be designed to influence global leaders prior to their preparation for the UN Climate Summit in Paris (COP 21) (*The Christian Science Monitor*, 18 June 2015).

Obama's climate change legacy

During the 2008 presidential campaign both major-party candidates (i.e. Obama and McCain) had backed "far more activist national and international approaches to global warming than [had] the Bush administration" (*The Christian Science Monitor*, 13 August 2008). With the inauguration of Barrack Obama as the new president in January 2009 there was a widespread expectation of "a more aggressive federal approach to confronting global warming" (*The Christian Science Monitor*, 13 August 2008). President Obama could not meet this expectation due to a variety of reasons. **First**, Republican opposition to climate change initiatives has been the most basic reason for his failure to pass any climate change legislation. **Second**, the cost of climate change initiatives has been a concern for many. In his speech to the Congress in February 2009, shortly after his inauguration, President Obama urged the Congress to try "to truly transform our economy, protect our security, and save our planet from the ravages of climate change" (*USA Today*, 2 March 2009). In response, people have been "wondering how much this is going to cost," said Peter Barnes, author of *Who Owns the Sky?: Our Common Assets and the Future of Capitalism*. "There are approaches out there that could benefit many people or ones that could benefit just few" (*USA Today*, 2 March 2009). **Third**, some have questioned if Obama's change of bold campaign rhetoric into a more pragmatic and cautious approach after taking over office was partly responsible for lack of a more significant progress on climate change issues. "President Obama's most constant refrain in word and action has been 'the perfect is the enemy of good'" (*The Christian Science Monitor*, 20 December 2009). The 2009 Copenhagen Accord (COP 15) is an example of his compromise on climate change issues. Prior to COP 15 he stressed that without firm commitments to a timeline for reducing carbon emissions "any agreement would be empty words on a page". At the end of the conference he just fulfilled his own prophecy by praising this agreement as an "important breakthrough", though there were no short-term or mid-term goals and no mechanism for enforcement (*The Christian Science Monitor,* 20 December 2009). In short, COP 15 proved to be a disappointment.

Despite these initial shortcomings, some of his ardent supporters, such as David Doniger of the Natural Resources Defense Council, have hailed Obama's subsequent record on climate change policies as "terrific" (Tollefson 2016). Former New York Mayor Michael Bloomberg echoed the same sentiment: "Obama did take major steps to reduce carbon emissions, including setting higher fuel efficiency standards for cars and trucks and adopting tighter controls on mercury emissions, which will help to close the dirtiest coal-fired power plants" (*The Christian Science Monitor,* 3 November 2012). In 2009, Obama signed economic-stimulus legislation that included nearly $37 billion for clean-energy research and development (R & D) at the Department of Energy. Further, "with failing car companies seeking a federal bailout, the Obama administration proposed higher fuel efficiency requirements and the first greenhouse gas standards for passenger vehicles (Tollefson 2016). To Obama's credit, even after his climate bill failed in 2010 he used existing laws to issue regulations that curbed greenhouse gas emissions bolstering energy-efficiency standards and expanded energy R & D programs (Tollefson 2016). The Republicans cried foul as it was evident from political backlash on Capitol Hill. The president's high-profile speech on climate change in late June 2013 sent a message to the Republican members of Congress that "their opposition to climate regulations . . . isn't going to stand in the way of administrative action" (*USA Today,* 25 June 2013).

Further to Obama's credit, in 2014, he managed to find a common ground on climate change with the Chinese President Xi Jinping. "As leaders of the two biggest greenhouse gas emitters, responsible for nearly 40 percent of global emissions, they have been under heavy international pressure to lead by example in the run-up to . . . the [2015] climate treaty summit in Paris" (COP 21) . . . "Xi pledged that China's CO_2 emissions would peak around 2030—the first time that Beijing has set such a target. Obama promised that by 2025 the US will have reduced its emissions by 23-26 percent of 2005 levels—twice as much as Washington had previously offered" (*The Christian Science Monitor*, 12 November 2014). The year 2015 proved to be an eventful year for Obama's legacy on climate change and environment. In November 2015, Obama announced his bold rejection of the +1,200 mile long Keystone XL Pipeline that was supposed to carry carbon-heavy crude oil from Alberta (Canada) tar sands to Texas refineries. In his announcement of the decision to reject the pipeline the President linked it to potential climate change impacts [of carbon dioxide emissions of petroleum]: America is now a global leader when it comes to taking serious action to climate change, and frankly, approving this project would have undercut that leadership" (Whitehouse 2016: https://www.whitehouse.gov/the-press-office/2015/11/06/statement-president-keystone-xl-pipeline/ . . . retrieved on 10 October 2016).

Among all of his climate change achievements undoubtedly he will be remembered most fondly for his bold leadership role in securing the global climate change agreement at the Paris Climate Summit (COP 21: December 2015), "the first such accord committing nearly every country to reducing greenhouse gas emissions" (*The New York Times*, 8 September 2016). It took nearly six years of post-Copenhagen (i.e. post COP 15) negotiations and diplomacy to pass this historic accord when 195 countries adopted the first-ever universal, legally binding global climate deal. "The agreement sets out a global action plan to put the world on track to avoid dangerous climate change by limiting global warming to well below 2°C (3.6°F). The agreement is due to enter into force in 2020" (http://ec.europa.eu/clima/policies/international/negotiations/paris/index_en.htm/ . . . retrieved on 15 October 2016). Obama's success at COP 21 may be attributed partly to his diplomatic skills and partly to sheer hard work by his negotiating team headed by his Secretary of State John Kerry. To Obama's credit, he brought on board such unlikely partners as President Xi Jinping of China and Prime Minister Modi of India. Among countries which are not signatories to the Kyoto Protocol China and India are two of the major contributors of carbon dioxide emissions.

Despite such landmark international achievements in climate change negotiations, paradoxically his domestic legacy on climate change is significantly diminished due to his inability to compromise with the Republicans and to some extent with some of his own party members who do not agree with his climate change policies. For nearly eight years in office, Mr Obama has claimed that "a majority of Americans have come to believe 'that climate change is real, that it's important and we should do something about it.' He enacted rules to cut planet-heating emissions across much of the United States economy, from cars to coal plants." Yet, in the final analysis, "while climate change has played to Mr Obama's highest ideals—critics would call them messianic impulses—it has also exposed his weaknesses, namely an inability to forge consensus . . . on a problem that demands a bipartisan response" (*The New York Times*, 8 September 2016).

Letters to the editor

Differences between main news reports and letters

So far we have analyzed 47 news reports on climate change in three newspapers containing about 35,000 words. The bulk of this text has been written by professional journalists (correspondents, staff reporters, editors and other newspaper staff), who specialize in communicating substantive concepts (such as political news, scientific and technical news and others) in plain language for the general audience. A significant portion of such text includes direct quotes of politicians, scientists and other experts, activists and other spokespersons. In contrast to the text of most of the newspaper reports, letters to the editor provide direct opinions of readers. As data in Table 8.8 indicate, our sample of newspapers included 14 reports (column 1) which published 35 letters to the editor, containing 4,646 words (column 2). This amounted to 13% of 39,621 words analyzed; the remaining 87% accounted for the main reports. Despite limited space allocated for letters by newspapers, contents of most of the letters appear to be refreshing because these reflect direct public discourse on certain issues beyond some of the standard news frames.

Who wrote those letters?

"Letters to the editor are among the most widely read features in any newspaper or magazine. They allow [the reader] to reach a large audience" . . . People may write a letter to the editor "just to 'vent', or to support or criticize a certain action or policy" . . . (Jenette Nagy, Community Tool Box 2016: http://ctb.ku.edu/en/table-of-contents/advocacy/direct-action/letters-to-editor/main/ . . . retrieved on 16 October 2016). The majority of the letters in our sample were written by people identified only by their names and short address (such as a town or a city and the state). Some of the authors who indicated their affiliations included a U.S. senator, an Ambassador, several university professors and post-doctoral fellows, medical doctors and medical researchers, and administrators in charge of several climate change activist organizations. Table 7.8 (column 2) highlights the main contents of the letters. The following are their major themes.

Debate on global warming hypothesis

This was the central theme of several letters. In a letter to the *New York Times* editor on 14 May 2014, the writer/author pointed out that in the North Carolina Congressional debate all Republican candidates denied the fact that climate change illustrated "our failure to ask the right question." The author challenged: "On what climate studies or data is your belief based? This will help voters judge whether this is an informed opinion or simply a surrender to ignorance?" (*The New York Times*, 14 May 2014). In an earlier letter (in 2011) the author had similar criticisms of the Republicans for not accepting the global warming hypothesis. Commenting on a Sunday Review by the *New York Times* on a topic entitled "Where Did the Global Warming Go?" the author criticized that it was a "shameful fact that all Republican presidential candidates [of 2012 presidential election] deny that man-made climate change is a serious problem . . . America's politicians are virtually alone in the world in failing to act on climate change" (*The New York Times*, 24 October 2011). Countering the latter position, a climate change denier presented a forceful argument against the global warming hypothesis. He summed it up as follows: "Your article concludes that global warming agnosticism is mostly an American thing. I disagree. Around the world, the opinion that global warming is a clear and present danger is much diminished. There is a realization that the case for global warming was uncertain at best, and certainly greatly exaggerated. Global warming remains a hypothesis. At least in this one instance, the United States showed itself more prudent, and rightly more skeptical, than many. Politics has always been the plague of science" (*The New York Times*, 24 October 2011). Similarly, another letter questioning the anthropogenic global warming hypothesis wrote: "They all seem to treat the science of human-caused warming of the planet as an assumption, rather than a question—even though there is little evidence of warming over the past 15 years." (*The Christian Science Monitor*, 18 October 2014).

Climate change impacts

Climate change impacts have been publicized heavily in the news media, based largely on IPCC projections and direct field evidence (often displayed in TV or published in newspapers as photographs). Sea level rise and melting of Greenland and Antarctic ice are two familiar examples. Perhaps, based on IPCC projections on sea level rise, one author wrote: "In recent months the scientific community has acknowledged that the worst-case scenarios are increasingly likely, and that past projections were far too conservative. Yes, we will likely see three feet of sea-level rise by the end of this century. Yes, even six feet cannot be ruled out" (*The New York Times*, 8 May 2014). Perhaps, referring to pictures of melting ice, the same letter added: "I hope that people will start to listen now that climate change is literally right in front of our eyes, and the melting of the ice in Greenland and Antarctica is already occurring much faster than initially projected" (*The New York Times*, 8 May 2014). Referring again to sea level rise and melting glaciers, another author attempted to implicate the Congress for inaction on climate change: "Experts warn that changes in the oceans will cause mass extinctions and that the situation is more dire than previously thought . . . The melting ice caps are visible, yet many in Congress won't admit it or rule on the remedies" (*USA Today*, 5 July 2011). Often politicians and the general public have a tendency to link any major climatic disaster almost automatically to climate change. Hurricane Sandy (October/November 2012) is a good example. Referring to an earlier report by Nicholas Kristof (*The New York Times*, 31 October 2012) with the caption "Will climate change get some respect now?" which cautioned readers that it was not possible to link a single disaster (event) to climate change, the author of a follow-up letter (actually a physical scientist/professor) reinforced public perception by implying that hurricanes like Sandy and Katrina were supposed to change political stance on climate change. "Until we start thinking beyond ourselves climate change will remain the Rodney Dangerfield of problems" (*The New York Times*, 2 November 2012).

Perhaps, one of the most passionate letters stressing the threat of climate change impacts was written by a medical doctor (a 1995 Nobel Peace Prize winner and director of the Harvard Center for Health and Global Environment), who compared the threat of climate change to that of nuclear war: "By giving its 2007 Peace Prize to Al Gore and the Intergovernmental Panel on Climate Change, the Nobel Committee has warned us of a new global threat, in magnitude and in the duration of its impacts comparable to nuclear war, but one that is harder for most people to recognize." (*The New York Times*, 16 October 2007). The author went one step further by positing that "Global climate change is Armageddon in slow motion, dangerously altering the atmosphere, land, oceans and life on Earth, including human life, in incremental steps" (*The New York Times*, 16 October 2007).

Public support for reduction of carbon pollution

In responding to a *New York Times* Sunday Review (16 October 2011) which noted that "Europe, Australia, China, India and Brazil are all moving ahead with policies to reduce heat-trapping pollution, while most American politicians duck the issue or actively question the reality of the problem," the author of a letter provided survey data indicating that the U.S. citizens were ahead of their politicians. He pointed out that a CNN poll indicated that, unlike politicians, 71 percent of the U.S, citizens believed that "the Environmental Protection Agency should be allowed to do its job of protecting us from dangerous carbon pollution" (*The New York Times*, 24 October 2011). The letter provided additional data: "Yes, American cars and homes are larger, on average, than those in Europe, but we support stricter energy efficiency standards by 73 percent to 26 percent. And 76 percent of Americans trust scientists for information about global warming, while only 37 percent trust their members of Congress, according to researchers at Yale and George Mason Universities" (*The New York Times*, 24 October 2011).

In a post-COP 21 development (i.e. immediately following the Paris Climate Summit), the U.S. Supreme Court rejected an Obama plan to regulate emissions from coal-fired plants (*The New York Times*, 12 February 2016). Criticizing this decision, the author of a letter vented that it "is a slap in the face to all Americans working to reverse global warming . . . This program is the foundation of our nation's first major action on climate. To freeze it before any case has reached the court demonstrates to the world that our judiciary has no respect for the accord reached in Paris by nations throughout the world . . . Climate change must not be a partisan issue, and the conservative members of the court should not politicize it." (*The New York Times*, 12 February 2016). Another letter in the same report had a similarly scathing indictment of the Supreme Court decision: "Notwithstanding that a unanimous Court of Appeals for the District of Columbia Circuit refused to block the Clean Power Plan from taking effect, five members of the Supreme Court, in an unprecedented decision without waiting until it formally heard the case, issued an order temporarily blocking the law." (*The New York Times*, 12 February 2016). "The coal industry should consider itself fortunate that it could do without the services of lobbyists, since five members of the Supreme Court will accomplish its bidding. History will not favorably judge members of the court who abandon sound legal reasoning and base their decision on right-wing politics." (*The New York Times*, 12 February 2016).

Carbon tax

Carbon tax (carbon fee) is one of the economic tools for reducing carbon emissions. As a climate change mitigation measure carbon tax has drawn attention from at least two climate scientists (researchers in atmospheric science) and a climate change activist. One of the scientists suggested that "A national carbon fee, if returned in its entirety as an equal dividend to consumers, would actually bring economic benefits. With one fair, transparent and comprehensive market correction, you reduce emissions, drive the transition to renewables, create jobs and stimulate consumer spending . . . Even Big Oil has begun lobbying for a carbon fee. These companies reason that the world will soon do something about global warming, and out of the available options only a carbon fee provides a predictable framework for their future development (presumably into green energy). The majority of Americans and economists also support a carbon fee." (*The New York Times*, 1 October, 2015). Earlier, another climate scientist echoed similar sentiments: "At some point, soon I hope, a revenue-neutral carbon tax will become politically possible. Revenue would be returned to taxpayers, rewarding those who emit less carbon and stimulating change throughout our economy and infrastructure" (*The New York Times*, 8 May 2014).

A climate change advocacy group (representing a Washington State-based organization) suggested that the quickest and most efficient solution for reducing carbon emissions in USA was "to turn to the free market and enact a carbon fee." (*The New York Times*, 1 October 2015). He pointed out that "more than 40 countries have adopted some form of carbon pricing". Praising British Columbia's successful carbon tax, the author continued: "British Columbia's revenue-neutral carbon fee has reduced fossil fuel emissions by 16 percent, while emissions in the rest of that country (Canada) have risen 3 percent. Meantime, British Columbia's growth outperforms Canada's". (*The New York Times*, 1 October 2015). In contrast to British Columbia's positive experience with carbon tax, Australia's experiment with it has not been so popular: "American readers may get the impression that Australia's introduction of a carbon tax plan is indicative of a national embrace of environmentally progressive policies. In fact, the tax has been nothing if not controversial. The main opposition party, backed by an increasingly hysterical right-wing media, has virulently opposed the tax. As a result, the current government, a fragile coalition, has seen its political capital plummet. The Labor Party now faces likely defeat at the next election. In these overheated times, American and Australian politics have this in common: take a stance to slow global warming and risk political suicide". (*The New York Times,* 24 October 2011).

Keystone XL Pipeline

The Keystone XL was proposed as an extremely complex pipeline system consisting of several components of pre-existing pipelines and new pipelines in both USA and Canada. The older pipelines were supposed to be upgraded with larger diameter to be consistent with the newer components. The project had four distinct phases. Three of the phases refer to existing pipeline systems in USA. The fourth phase was known as the Keystone XL pipeline that was "an expansion of Trans-Canada's existing system to funnel oil from Alberta's tar sand to refineries in the United States" (http://phys.org/news/2015-11-key-facts-controversial-keystone-xl.html/ . . . retrieved on 21 October 2016). The application for a permit to build the pipeline was rejected by President Obama twice in a row. For the first time he was forced to reject the permit in January 2012 under pressure from his Republican opponents in the Congress to make a quick decision before the Presidential election in November 2012 (*The Washington Post*, 18 January 2012). The Obama administration allowed Trans-Canada to reapply for the permit, but in November 2015 Obama rejected the proposal, "ending a seven year review that had become a symbol of the debate over his climate policies" (*The New York Times*, 6 November 2015).

From the very beginning of the proposal the U.S. public was divided in their opinions on the project. Pro-oil lobbies and their Republican supporters in the Congress as well as some of the labor unions working in the pipeline industry supported the proposal, stressing potential job creation by the new construction project. Energy independence from foreign oil (i.e. from outside the North American market) was perhaps the most important incentive for the supporters of the project. In the final analysis, however, adverse environmental impacts and its potential contribution to climate change killed the project. "Tar sand oil must be dug up and essentially melted with steaming hot water before it can be refined. This means more fossil fuels need to be burned as part of the extraction process, which further contributes to climate change. It also results in huge lakes of polluted water and strip-mining of millions of acres of once-pristine boreal forests" (http://phys.org/news/2015-11-key-facts-controversial-keystone-xl.html/ . . . retrieved on 21 October 2016).

In 2012, when Obama rejected the Keystone XL proposal for the first time, the *New York Times* published a column (by Joe Nocera) on "Poison Politics of Keystone XL," reviewing pros and cons of the project (*The New York Times*, 6 February 2012). This review attracted four letters to the editor, which were published on 11 February 2012. Two of the letters were supportive of the President's decision and highly critical of pro-Keystone argument. One letter pointed out that "the job projections were discovered to be highly inflated; State Department estimates barely break 5,000". The letter further pointed out that "many Canadians oppose a tar sand pipeline as it is destroying large sections of the environment. Mix this with a dozen leaks that Keystone I has had in a single year, extremely high refining costs, and a route that puts American food and water supplies at increased risk, and it's hard to find a reason to like Keystone XL" (*The New York Times*, 11 February 2015). Another letter expressed similar views: "The only way to avoid dangerous, irreversible climate change is by reducing fossil fuel dependence at every opportunity, not by investing billions in fossil fuels development." The third letter was a dissenting voice: "Blocking the Keystone project will not reduce American dependence on imported oil, preclude development of Canada's oil sands, or have any meaningful impact on climate change. What it will do is squander the benefits of having huge oil reserves just across the border under the control of a true and trusted ally and put the seas at risk from super-tankers that will carry that oil to Asia. That's flawed policy on both security and environmental grounds" (*The New York Times*, 11 February 2015). The fourth letter (from a former U.S. ambassador to Canada) commented on the politics of the Keystone debate: "President Obama's decision to reject Trans-Canada's application had nothing to do with either energy or environmental concerns but was based entirely on electoral politics . . . which is not to say that the Republicans handled the issue responsibly [before the presidential election].

Pope's encyclical on climate change

Following Pope's encyclical on climate change, Republican presidential candidate Rick Santorum, a devout practicing Catholic, did not approve pope's message on climate change. He said: "The pope should stay out of the climate change debate". Mr Santorum added: "We probably are better off leaving science to the scientists." One of the letters to the editor, published in *USA Today*, criticized Santorum for his comments: "If that's the case, Santorum and other Republicans need to follow his advice. Scientists have spoken on the issue. The debate is really over. Almost unanimously, climate scientists agree that climate change is occurring and that man is a contributing factor. Leaders would serve better by addressing this issue rather than denying it." (*USA*

Today, 18 June 2015). The *USA Today* also followed this issue by asking its tweet followers "What they thought about Pope Francis weighing in on climate change". The following were the samples of tweet responses (*USA Today*, 18 June 2015):

- "The issue of climate change is of such importance that all support is needed. Efforts by President Obama are also very good."
- "The left will use the pope as a metaphorical club in the climate debate, but the real impact is a damaged, politicized papacy".
- "It's not political. It is the future of our world. The politicians made it political pandering to their campaign donors".

CONCLUDING COMMENTS

Sampling limitations

Media scholars have wrestled with the competing issues of adequate sample size of newspaper reports versus the appropriate sampling method for obtaining them (Lacy and others 2001). Random sampling may be the preferred method of obtaining the minimum numbers of reports required for an adequate sample size, but this method is contingent upon the following statistical characteristics of the original data: (a) original population size can be determined (delineated) clearly, (b) original population size is large enough, and (c) the population is normally distributed (see websites for relevant information and/or scholarly articles on "random sampling" and "normal distribution"). While digital revolution has opened up access to huge numbers of electronic copies of newspaper reports, we have found out during our initial search that it is difficult to determine the true size of the population (i.e. the total numbers of reports on a given topic, such as climate change politics in USA). Our initial search through the LexisNexis retrieval system (database) yielded at least 1,000 news reports. We also noticed that related keywords, such as climate change, greenhouse gas emissions, emission reductions, climate change mitigation and adaptation, and similar other terms, also yielded large numbers of reports. At the beginning of this chapter we have commented briefly on our sampling method. We could not employ a rigorous random sampling method because we ended up with only 61 U.S. newspaper reports following our screening method. A sample of only 61 reports was too small for drawing another blind sub-sample by using a random number table (i.e. the random sampling method). Thus, we are not entirely certain how to characterize our sampling method. At best, our screening method amounted to a type of "stratified sample" (again, refer to websites for this type of sampling method). At worst, we ended up with an arbitrary sample. The main shortcoming of this arbitrary sample is that we have missed many relevant reports on several climate change-related topics. In particular, our sample does not include any specific reports on COP-15 (at Copenhagen), COP 17 (at Durban) and COP 20 (at Lima). Similarly, we have missed some of the relevant reports on the Keystone XL pipeline, such as the *New York Times* report on Obama's first rejection of the pipeline in January 2012 (we have retrieved this particular report through a different search).

Context issues

By focusing solely on newspaper reports we have encountered not only the problem of inadequate sampling but, perhaps more importantly, the issues of lack of contexts. Without adequate background on a topic (i.e. its context) it is often difficult to follow some of the reports, which provide at best some of the scattered glimpses of information. A number of reports on the Keystone XL pipeline provide a good example of inadequate contexts. The Keystone XL is a complex topic as it involves different physical components and different phases of proposed construction. None of our sample reports provided these details and it was difficult to follow them. To remedy this problem, we have provided some background information on the project based on a website (see above for the section on the Keystone XL pipeline).

Future research

Notwithstanding the preceding shortcomings, we have been able to prepare a comprehensive research report (i.e. this chapter) based largely on newspaper reports. We have relied almost entirely on digital copies of newspaper reports, which were supplemented with some of the website documents, including government and non-government blogs.

We have also used information from a limited number of peer-reviewed scholarly journal articles and textbooks. We accessed these articles and books sometimes directly from websites (depending upon availability) and most of the time through the website of the Murphy Library of the University of Wisconsin-La Crosse (I have access to digital copies of these articles and books as an emeritus professor).

Digital revolution has truly transformed media research. Based on our experience of research on climate change politics in USA, we have learned a valuable lesson for future research on a similar topic. It is more appropriate to construct a theoretical outline of an article (or a chapter) at the very beginning of the project. Then retrieve relevant information simultaneously from at least three digital resources: electronic newspaper reports, website blogs, and scholarly journal articles. The final outline of the article/chapter may then be revised based on the progress of the article following thorough reviews of relevant information from different sources.

References

EPA (Environmental Protection Agency) 2016a. *Draft U.S. Greenhouse Gas Inventory Report: 1990-2014.* (https://www3.epa.gov/climatechange/science/indicators/ghg/us-ghg-emissions.html . . . retrieved on 15 April 2016.

EPA. 2016b. *Climate Change Indicators in the United States: U.S. and Global Temperature.* (https://www.3.epa.gov/climatechange/science/indicators/weather-climate/temperature.html . . . retrieved on 25 April 2016).

Fischer, D. 2013. "Dark Money Funds Climate Change Denial Effort". *Scientific American.* Web document: http://www.scientificamerican.com/article/dark-money-funds-climate-change-denial-effort/ . . . retrieved on 4 October 2016.

Fletcher, A.I. 2009. "Clearing the Air: The Contributions of Frame Analysis to Understanding Climate Policy in the United States." *Environmental Politics* 18 (5): 800-816.

Houghton, J. 2004. *Global Warming: The Complete Briefing,* Third Edition. Cambridge, UK: Cambridge University Press.

Hurley, L. and Volcovici, V. 2016. "U.S. Supreme Court Blocks Obama's Clean Power Plan". *Scientific American* (web article, 9 February 2016). www.scientificamerican.com/article/u-s-supreme-court-blocks-obama-s-clean-power-plan/ . . . retrieved on 21 August 2016.

IER (Institute for Energy Resources). 2014. *Hard Facts: An Energy Primer*. Second Edition. Web document (instituteforenergyresearch. org/ . . . retrieved on 20 August 2016).

Kolmes, S.A. and Butkus, R.A. 2007. "Science, Religion and Climate Change". *Science* 316 (5824, April 2007): 540.

Lacy, S. Riffle, D., Stoddard, S., Martin, H. and Chang, K-K. 2001. "Sample Size for Newspaper Content Analysis in Multi-Year Studies". *Journalism and Mass Communication Quarterly* 78 (4): 408-417.

NASA. 2016. *Global Climate Change: Vital Signs of the Planet* (climate. nasa.gov/evidence/ . . . retrieved on 24 April 2016).

NCA (*National Climatice Assessment*). 2014. (nca2014.global change. gov/report/our-changing-climate/recent-us-temperature-trends . . . retrieved on 26 April 2016).

Tollefson, J. 2016. Obama's Science Legacy: Climate (Policy) Hots Up." *Nature* (News) 536 (7617): not paginated web document.

Table 8.1   Greenhouse gas Emissions in USA: 1990-2014

Year	Greenhouse gas emissions (CO_2 equivalent) Million metric tons*
1990	6,381
1995	6,739
2000	7,302
2005	7,429
2010	7,010
2014	6,873

*   1,000 million metric tons = 1 Gt (gigatonnes). Thus, 6,381 million metric tons = 6.38 Gt.

Source: Prepared by the first author based on original data in EPA (Environmental Protection Agency) 2016a. *Draft U.S. Greenhouse Gas Inventory Report: 1990-2014.*

(https://www3.epa.gov/climatechange/Downloads/ghgemissions/ . . . retrieved on 15 April 2016.

Table 8.2 Main Sources of Greenhouse Gas Emissions in USA, 2014

Types of greenhouse gases:	Emissions in million metric tons (% of total)
All greenhouse gases	**6,873** (100)
Carbon dioxide	5,564.3 (81)
Methane	707.9 (10)
Nitrous oxide	411.4 (6)
Others	206 (3)
Carbon dioxide emissions by fossil fuels: leading economic sectors:	**5,208.7** (94% of total carbon dioxide emissions)
Electricity generators	2,039.3 (39% of fossil fuels)
Transportation	1,737.4 (33)
Industries	814.3 (16)
Residential	345.1 (7)
Commercial	231.6 (4)
Others	52 (1)

Source: Prepared by the first author based on original data in EPA (Environmental Protection Agency) 2016a. *Draft U.S. Greenhouse Gas Inventory Report: 1990-2014.*

(https://www3.epa.gov/climatechange/science/indicators/ghg/us-ghg-emissions.html . . . retrieved on 15 April 2016).

Table 8.3 Average Annual Temperatures in USA*, 1901-2015

Decade	Average annual temperature** °F (change: °F per decade)	Average annual temperature °C (change: °C per decade)
1901-1910	51.54	10.86
1911-1920	51.25 (-0.29)	10.69 (-0.16)
1921-1930	51.94 (0.69)	11.08 (0.38)
1931-1940	52.63 (0.69)	11.46 (0.38)
1941-1950	52.00 (-0.63)	11.11 (-0.35)
1951-1960	52.17 (0.17)	11.21 (0.09)
1961-1970	51.71 (-0.46)	10.95 (-0.26)
1971-1980	51.75 (0.04)	10.97 (0.02)
1981-1990	52.43 (0.68)	11.35 (0.38)
1991-2000	52.47 (0.04)	11.37 (0.02)
2001-2010	53.25 (0.78)	11.81 (0.43)
2011-2015	53.57 (0.32 for 4 years)	11.98 (0.18)
First half: 1901-1960	**51.92**	**11.07**
Second half: 1961-2015	**52.93**	**11.41**
Average change per decade: 1901-2010	**0.17**	**0.10**

*   For 48 contiguous states excluding Alaska and Hawaii.

** Average annual temperature has been computed by averaging monthly temperatures for 12 months (January to December) for 10 years (for example, 1901-1910). Thus, data for each decade include 120 months (i.e. 12 months x 10 years).

Source: Prepared by the first author based on original data in NOAA (National Oceanic and Atmospheric Administration) 2016. *Climate at a Glance: Time Series.* (https://www.ncdc.noaa.gov/cag/time-series/us/ . . . retrieved on 20 April 2016).

Table 8.4   Average Seasonal Air Temperatures in USA*: 1901-2015

Decade	January** °F (°C)	April °F (°C)	July °F (°C)	October °F (°C)
1901-1910	30.52 (-0.82)	50.6 (10.33)	72.84 (22.69)	53.6 (12.00)
1911-1920	29.25 (-1.53)	50.36 (10.20)	73.14 (22.86)	53.4 (11.89)
11921-1930	29.48 (-1.40)	51.15 (10.64)	73.55 (23.08)	53.7 (12.06)
1931-1940	30.75 (-1.69)	50.98 (10.54)	75.07 (23.93)	54.9 (12.72)
1941-1950	30.14 (-1.03)	51.58 (10.88)	73.33 (22.96)	54.95 (12.75)
1951-1960	30.75 (-0.69)	51.17 (10.65)	73.93 (23.29)	54.35 (12.42)
1961-1970	28.94 (-1.70)	50.92 (10.51)	73.47 (23.04)	54.28 (12.38)
1971-1980	28.48 (-1.96)	50.74 (10.41)	73.55 (23.08)	53.82 (12.12)
1981-1990	31.14 (-0.48)	51.83 (11.02)	73.74 (23.19)	53.77 (12.09)
1991-2000	31.80 (-0.11)	51.08 (10.60)	73.49 (23.05)	54.35 (12.41)
2001-2010	32.46 (0.26)	52.38 (11.32)	74.65 (23.69)	54.34 (12.41)
2011-2015	32.34 (0.19)	52.25 (11.25)	74.87 (23.82)	55.32 (12.96)
First half: 1901-1960	**30.15 (-1.03)**	**50.97 (10.54)**	**73.64 (23.13)**	**54.15 (12.31)**
Second half: 1961-2015	**30.86 (-0.63)**	**51.53 (10.85)**	**73.96 (23.31)**	**54.31 (12.39)**

*   For 48 contiguous states excluding Alaska and Hawaii.

**   Months of January, April, July and October represent sample months for the winter, spring, summer and fall seasons, respectively. Data under each column are the average temperatures (in °F and °C) for a given month for ten years (such as for 1901-1910).

Source: Prepared by the first author based on original data in NOAA (National Oceanic and Atmospheric Administration) 2016. *Climate at a Glance: Time Series.* (https://www.ncdc.noaa.gov/cag/time-series/us/ . . . retrieved on 20 April 2016).

Table 8.5 Numbers of Reports on Climate Change in Some of the Leading Newspapers (based on *LexisNexis* Electronic Database)

Newspapers	Numbers of reports
The Age (Melbourne, Australia)	79
Sydney Morning Herald (Australia)	56
The Guardian (UK)	54
Australian Financial Review	44
The New York Times	34
The Mercury News (California)	28
Christian Science Monitor	27
McClatchy Tribune (non-restricted)	25
The Atlanta Journal-Constitution	23
The East Bay Times (California)	23
The Washington Post	23
USA Today	23

Source: *LexisNexis® Academic* (accessed through Murphy Library, University of Wisconsin-La Crosse . . . retrieved on 29 March 2016). Prepared by the first author based on discourse analysis of selected newspaper reports.

Table 8.6 Data on Lengths of Newspaper Reports on Climate Change-Related Issues

Newspaper (number of reports)	Total words analyzed	Average length of all reports (words)	Longest report (words)	Shortest report (words)	Median length of all reports (words)
*NYT (13)	11,584	891	1,239	300	1,056
CSM (21)	14,908	710	975	340	731
UST (12)	8,483	707	1,468	329	612

*   NYT: *The New York Times*; CSM: *The Christian Science Monitor*; UST: *USA Today*

Source: Prepared by the first author based on discourse analysis of selected newspaper reports.

Table 8.7 Discourse Analysis of Newspaper Reports on Climate Change-Related Issues

Nature of data	Total words (% of total words)		
Climate change-related themes	*NYT	CSM	UST
Total numbers of words analyzed [numbers of reports analyzed]	11,584 [13]	14,908 [21]	8,483 [12]
Global warming hypothesis and climate change skeptics/deniers	1,404 (12%)	1,577 (11%)	758 [8.94]
Climate change impacts	938 (8%)	948 (6%)	534 [6.29]
Climate change mitigation and adaptation	1,941 (17%)	2,621 (18%)	703 [8.29]
Climate change politics and policies	5,099 (49%)	7,166 (48%)	4,362 (51.42%)
Others	1,602 (14%)	2,596 (17%)	2,126 (25.06%

* NYT: *The New York Times*; CSM: *The Christian Science Monitor*; UST: *USA Today*

Source: Prepared by the author based on discourse analysis of selected newspaper reports.

Table 8.8 Letters to the Editor

Newspaper (date)	Letter # (words), main theme(s) of each letter
The New York Times (12 February 2016)	L-1 (126): Supreme Court decision to reject Obama Plan to regulate emissions from coal-fired plants. L-2 (96): Same as L-1
The New York Times (1 October 2015)	L-3 (187):Carbon fee/tax L-4 (155): Need for corporate climate policies L-5 (176): Urges Congress to act on carbon pricing
The New York Times (8 May 2014)	L-6 (169): Revenue neutral carbon tax L-7 (119): Climate change impacts, including sea level rise and melting ice L-8 (130): Same as L-7 L-9 (114): Leadership for data-based climate change
The New York Times (27 November 2012)	L-10 (118): Criticizes government subsidies to rebuild in disaster-prone lands L-11 (138): Post-Sandy coastal reconstruction; climate change and energy security
The New York Times (2 November 2012)	L-12 (107): Post-Sandy bipartisanship L-13 (129): Politics of climate change denial L-14 (67): Post-Sandy gasoline shortage
The New York Times (11 February 2012)	L-15 (136): Adverse impacts of Keystone XL pipeline L-16 (143): Failure to utilize benefits of Keystone pipeline L-17 (95): Justifying rejection of Keystone pipeline L-18 (79): Politics of Keystone pipeline
The New York Times (24 October 2011)	L-19 (193): Criticizes Americans/Republicans for not accepting climate change L-20 (111): Criticizes Americans/Republicans for not accepting climate change L-21 (81): Uncertainty with global warming hypothesis L-22 (133): Criticizes lack of federal leadership on climate change L-23 (99): Unpopular Australian carbon tax
The New York Times (16 October 2007)	L-24 (220): Nobel Prize for climate change (alerting us about the threat of climate change)
The New York Times (20 September 1995)	L-25 (222): Criticizes coal and oil lobby and politicians for blocking climate change legislations
The Christian Science Monitor (10 January 2015)	L-26 (168): Include Taiwan in climate change dialogue L-27 (144): COP 20 (Lima): More checks needed for emissions
The Christian Science Monitor (18 October 2014)	L-28 (195): Renewable energy is not the answer; questioning global warming hypothesis
USA Today (18 June 2015)	L-29 (78): Tweets on Pope Francis's encyclical L-30 (97): Criticisms of politicians regarding Pope's encyclical
USA Today (5 July 2015)	L-31 (136): Fuel economy car L-32 (93): Experts warn of climate change impacts
USA Today (16 March 2010)	L-33 (149): Broad strategy for global climate treaty L-34 (87): Criticism of climate scientists L-35 (156): Critical of UN/developing countries for climate justice demand

Source: Prepared by the first author based on discourse analysis of letters to the editor.

Chapter 9

CLIMATE CHANGE
POLICIES IN CANADA

Geographical settings of Canada

In this chapter we will discuss the impact political opinion has had on Canada's reputation on Climate change on the world scene. One might describe Canada from various perspectives. It is the second largest country in the world, with only Russia being larger in area. This large landmass is bounded by three oceans. The majority of the population lives within 160 km (100 miles) from its southern border with the United States. The largest population density is in Southern Ontario and Southern Quebec. The spring and summer of 2011 saw unusual flooding in Manitoba and Quebec. Extreme heat waves are common in Toronto and drought conditions in Alberta are becoming more frequent. The occurrences of extreme weather conditions are not unheard of in Canada but they are becoming more frequent. Canada has four climate zones that include, dry climates, humid mesothermal climates, humid microthermal climates, and polar climates.

According to Gall and Antone (2012), Canada compares poorly with its peers on measures that address climate change. The provinces of Ontario and Alberta have the highest emissions among the provinces and territories. The enormous size of Canada and hence the long distances between products and their markets, are a major contributing factor in greenhouse gas emissions. Managing the extreme weather conditions in Canada also contribute to greenhouse gasses. The burning of fossil fuels in the winter, and coal generated electricity for air conditioning in the summer also add greenhouse gas emissions by Canadians every day.

Canada's positions on the Kyoto Protocol

The Canadian Federal Government, under the leadership of former Prime Minister Jean Chrétien (1993-2003), was one of the first countries in the world to ratify the Kyoto Protocol in 2002. This was seen as a forward step toward the control of greenhouse gas emissions and reductions of the risk of global climate change. The Kyoto Protocol led to the development of fifteen acts and seven federal regulations that relate directly to climate change. Due to the declining dollar in the intervening decade, Canada was forced to rely on the export of Oil from the Alberta tar sands, and coal to boost trade revenues. This pressure resulted in the Federal Government under Prime Minister Harper (2006-2015) to withdraw from the Kyoto Protocol in 2011.

Post-Keystone alternate pipelines

Canada ranks third in the world for oil reserves. Alberta's tar sands represent 97% of these reserves. In 2011 the TransCanada Pipeline Company, proposed the development of an oil pipeline to deliver heavy crude oil from Alberta's tar sands to refineries in Texas. This has led to lots of controversies in both countries. The Keystone XL project was rejected by the U.S. government in 2015. In the absence of the Keystone pipeline the Canadian government has an alternate plan to push another pipeline through the Rocky Mountains from Alberta to British Columbia. This pipeline would supply tankers headed for the Asian markets. There are also Western provinces that are proceeding with exploration for Shale Gas for local consumption. Quebec is among the provinces opposed to pipeline expansion. British Columbia is opposed to the mountain pipeline project because of its concerns for environmental impacts of such a project running through fragile environments and mountainous terrain. Quebec has a moratorium on shale gas exploration as well.

Canada's coal reserves

According to Gall and Antone (2012), Canada has significant deposits of coal. The largest of these deposits are found in Alberta, Saskatchewan and British Columbia. Smaller deposits are found in New Brunswick and Nova Scotia, and yet to be developed coal resources in Ontario, Yukon Territory, Newfoundland and Labrador, Northwest Territories, and Nunavut.

Federal emission regulation programs

Canada remains the only coal exporting country in the world. This may sound to be bleak, but there is some good news to report. The federal and provincial governments have been working on a series of programs to improve Canada's position on greenhouse gas emissions. In 1995 the federal government implemented the EcoAction Community Funding Program according to *Canada's Emissions Trends*. The federal government has improved emission standards on cars and trucks nationally. This improvement includes the addition of renewable fuels to both automobile gas and truck diesel fuels that came into effect at the end of 2010. The federal government has invested it the development of energy efficient buildings, vehicles and appliances.

British Columbia's emission reduction programs

In recent years many provinces have moved forward on the development of climate change strategies. A leader among them is British Columbia, especially for introducing carbon tax on gasoline. Consumers at gas pumps have paid this tax without any significant protests. Beginning in 2008 British Columbia became the first carbon neutral public sector in North America. Many projects that have been spawned by this position have made a significant impact on greenhouse gas emissions in the province and led to the development of a whole new sector of the economy. Green jobs are on the rise in British Columbia, and across Canada. Although the trend is promising, the lack of coordination of initiatives between provincial and federal governments has been the largest stumbling block to improving Canada's success in controlling greenhouse gas emissions.

Canada's climate change policies

Overview

As you may have noticed, the federal government in Canada has played a major role in forming Canada's policy on climate change. Some decisions have been made on the basis of short term financial goals, not taking into account their long-term impacts. During the last decade, Canadian public has become more aware of the danger of climate change. Starting in 2015, world oil prices have fallen. Financial impacts of our worldwide trade in oil have played a smaller role in our economic recovery than expected. Successive Conservative governments since 2006 had stressed the oil and gas sector of the economy. With falling revenues, public confidence in the government's plans eroded. These two factors and many others led to the change in the federal government in Canada in 2015.

In the next section, we will present the platforms of the federal political parties as the 2015 election approached. You will notice a wide difference between the various parties' views of Canada's future. Based on a blog-post by Environmental Defence (2015), entitled *Will Canada Finally Tackle the Climate Challenge?: An Updated comparison of Federal Parties' Positions on Climate Change*, we will explore the positions of the federal parties leading up to the 2015 federal election. We will compare their position statements on post-2020 carbon reduction, their main mechanism for reducing carbon pollution, tar sands specific pollution, and how they are planning to fund these programs.

Climate change policies by federal parties

Prior to getting into the details of each party's position let us compare them in general, using data from Table 9.1.

Let's look at each party's positions in some detail.

Conservatives: the party pledges to reduce carbon emissions by 30 percent of the 2005 levels by 2030. This represents a 14 percent reduction of emissions below the 1990 levels by 2030. This reduction pledge is the weakest of the G7 countries. There was no commitment on the part of the government to fund any of these measures. These plans totally neglect the impact of the tar sands project.

NDP: The NDP is committed to reducing Canada's green house gas emission to 34% below the 1990 levels by 2025. This target will keep global climate change to 2 degrees Celsius (3.6°F). Canada's carbon footprint will subsequently be reduced by 80% by 2050. The NDP is also committed to providing funding assistance to developing countries to aid in their green projects.

Green Party: The Green Party's position is almost twice as strong as that of the Conservative position. They are pledging 27% below 1990 levels by 2025. The greens are proposing a tax on all fossil fuels used in Canada. The funds are to be distributed equally among all Canadians. The price could begin at $50/ton of carbon dioxide and go up to $200/ton by 2030. About $500 million dollars a year will be earmarked for assisting the international Climate Change Fund.

Liberals: Their position remained vague at the time of the election. They maintained that they would limit climate change to 2 degrees Celsius (3.6°F) by 2050. When pressed for more details they stated that they would consult with the Provinces to develop an appropriate mechanism to achieve this target.

Trudeau Government's position on climate change

At the time of this writing the election was over in October 2015. The Liberals won a majority government. The New Prime Minister Trudeau went to the Paris summit and presented a very well received position on behalf of the Canadian public. According to Mike Blanchfield of the *Canadian Press*, the EU sees Canada as an important player in the world of Climate change (Blanchfield 2015). The fact that he has linked climate change and energy policy is a major step forward. Europe has identified renewable energy, innovation and research, as well as imports of liquefied natural gas as topics it wishes to discuss with Canada. The first shipment of liquefied natural gas is due to leave New Brunswick bound for Germany in 2020. In order to achieve this goal, Canada must solve its transportation issues getting the liquefied gas to an Atlantic port. The government has proposed a new pipeline that would transport the gas across Canada. The project has been labeled Energy East. The Mayors from greater Montreal have met and declared their opposition to the proposed pipeline. Prime Minister Trudeau has met with the provinces within the 90 day time frame that he proposed during the election. The final outcome of these meetings has yet to be disclosed as new Canadian policy.

References

Blanchfield, M. 2016. "Trudeau's Arrival Marks New Era on Climate Change, Energy: EU." *The Canadian Press* 01/21/2016 3:15 pm EST

Environmental Defence. 2015. *Will Canada Finally Tackle the Climate Challenge?: An Updated Comparison of Federal Parties' Positions on Climate Change.* Environmental Defence blog: http://www.equiterre.org/sites/fichiers/climatechange_politicalparties_octupdate-lg_final_oct_1st.pdf/ . . . retrieved on 26 October 2016.

Gall, S.B. and Antone, M.K. editors. 2012. "Overview of Current Environmental Issues, Canada". In *WorldMark Encyclopedia of U.S. and Canadian Environmental Issues.* Detroit: Gale Cenage Learning. http://link.vpl.ca/portal/Worldmark-encyclopedia-of-U.S.-and-Canadian/u9B0JXmXx3k/ . . . retrieved on 26 October 2016.

Table 9.1   Greenhouse Gas Emission Reduction Pledge at Paris Climate Summit (2015) by Canadian Federal Parties

Party	2025 target (relative to 1990)	Main policy mechanism	Includes Alberta tar sands?	*Commitment to adaptation financing?
Conservative Party	-6%	Sector by sector regulations	No	No
New Democratic Party	-34%	Cap-and-trade	Yes	Yes, but no amount cited
Liberal Party	Promises that Canada will do its part to limit global warming to 2°C (3.3°F)	Setting principles that provinces must meet	Depends on Alberta's policies	No, but critical of government for not committing
Green Party	-26.7%	Fee-and-dividend	Yes	Yes: $500 million/year

*   Adaptation financing involves providing financial assistance to help developing countries adapt to climate change.

Source: Environmental Defence 2015.

Chapter 10

CONCLUSION

Climate change in public discourse

In the 1970s, most of the climatology textbooks included a chapter on climate change—mostly the very last chapter. The role of greenhouse gases in global warming was treated as one of the factors of climate change among other factors, such as sunspot activity, earth's orbital changes, reversal of geomagnetic poles, and several other natural geophysical factors. Climate change was hardly covered in public discourse. The creation of IPCC (Intergovernmental Panel on Climate Change) by the United Nations in 1988 brought climate change in the forefront of public discourse. Based on long-term data on greenhouse gas emissions, simultaneous increases in average global temperatures and results of climate change simulations (models), the IPCC made alarming projections of significant increases in global average temperatures and their actual and potential impacts on the environment, such as sea level rise, increases in tropical cyclones (hurricanes), increases in river floods and droughts, melting of land glaciers and sea ice caps. In the last three decades, research on climate change has increased almost exponentially. Many universities have now undergraduate and graduate courses in global warming and climate change. So far, advances in new knowledge in climate change have been remarkable, but the main findings of this research have challenged the fossil fuel-based activities of the industrial societies. More specifically, the main culprit of climate change is greenhouse gas emissions, particularly by industrial and automobile emissions of carbon dioxide. The bulk of the carbon emission occurs from burning fossil fuels, such as petroleum (oil) and coal. This has political and economic implications. To protect their self-interests, powerful oil and coal lobbies oppose legislative measures to regulate greenhouse gas emissions. Most of the oil and coal lobbies back Republicans who, by proxy, oppose legislations for reducing greenhouse gas emissions, citing anthropogenic global warming denial hypothesis as a defense of their legislative actions. The Democrats, in contrast, are mostly supportive of legislative measures for curbing greenhouse gas emissions. In the process, the American society has been split into two camps with opposing positions on climate change. High-profile politicians like Al Gore and his millions of followers (climate change activists) have added some passions to the climate change debate. The 2007 Nobel Peace Prize to Al Gore and the IPCC marked the climax of recognition of climate change as a clear and present danger for the global environment. The news media (television and newspapers in particular) now cover climate change news extensively, often reporting on the contested claims and counter-claims of the global warming hypothesis and various types of climate change impacts. On balance, it seems that most of the news coverage deals with information on greenhouse gases, global warming, and climate change impacts. Climate change denial hypothesis has drawn very little traction with the media. In short, the IPCC has transformed media and public discourse on global warming and climate change in a fundamental way.

Climate change science versus social amplification of risk

One of the problems with the media and public discourse on climate change is that often it is difficult for the general audience to grasp the main issues of climate change. Both the nature of climate change science and the media discourse are responsible for this confusion. Unlike meteorology (physics of the atmosphere), which is governed by a set of physical laws and mathematical equations, climate change is subject to interpretations, especially based on models. Models have parameters and inputs and the latter can be changed for experimental purposes. In particular, the cause and effect relationships are not necessarily indisputable. In the process of media discourse on various aspects of climate change some of the nuances are lost, because the media are expected to simplify complex scientific information for the lay people (Reckelhoff-Dangel and Petersen 2007). The adverse impact of such inadequate information or misinformation is that not only the lay people but also some of the high-powered politicians often assume uncritically (i.e. without any questions) cause-and-effect relationships between climate change and some of the natural hazards and disasters as a matter of fact. One remarkable example of this is a claim by several high-profile politicians in USA that hurricanes Katrina and Sandy were the results of global warming and climate change. The media discourse in USA is replete with reports on how many Republican leaders disputed a direct link between these storms and climate change whereas most of the Democrats insisted that there was a clear link. The truth lies perhaps between these two competing positions. Scientific research suggests that it is difficult to relate individual storms to climate change, although global warming and resulting warming of ocean water most likely played an important role. Similarly, droughts and forest fires in California have been linked to climate change by many scientists, while others (not necessarily climate change deniers) have questioned such a direct link.

The language of communication is very important in explaining climate change impacts. A classic example is provided by cyclone frequency in the Bay of Bengal. If you ask a lay person or to someone who is not familiar with research findings, he or she will almost invariably say that cyclone frequency in the Bay of Bengal has increased recently. Most likely he or she has been influenced by extensive media coverage of hurricanes/cyclones in different parts of the world, but this is a false answer. In chapter 6 we have provided data indicating that cyclone frequencies in the Bay of Bengal have decreased recently but the frequencies of high-magnitude tropical cyclones (hurricanes) have increased. This is a critical example of the importance of accuracy in popular science communication. One should be circumspect, i.e. very careful, before making a claim of direct relationship between climate change and a given climatic hazard. For another example, if anyone is so certain about a direct relationship between global warming and Atlantic hurricanes, can he or she explain why there was no landfall of a major hurricane in Florida in last 12 years despite significant spikes in global warming during the same period? In the absence of irrefutable climatic data, media and public discourse is replete with similarly incomplete, incoherent and often exaggerated information on climate change impacts. That is not called climate change denial. In social science research, this type of socially-constructed information is called social amplification of risk (Kasperson and others 1988).

What have we learned from this study?

It should be clear from the preceding analysis that the current media discourse on climate change is subject to social amplification of risk. One of our goals in this book is to identify for the general audience at least two types of information on climate change: one dealing with verifiable scientific information, while others that are subject to social amplification of risk. Since the science of climate change is at its early stage of development there are some scientific claims-making in progress. To address this issue, we have used a social constructionist approach to organize this book, as we indicated in the introduction (Hannigan 1996 and 2006). In this approach, chapters on greenhouse gases (chapter 2), what is global warming (chapters 3), who are polluting our atmosphere (chapter 4) and what is climate change (chapter 5) provide verifiable objective data, i.e. a set of scientific information and data. Assuming that you do not have a background in climate science, our hope is that the information in this book is adequate for your understanding of the basic concepts of climate change.

Another important lesson of this book is that some of the media discourse on climate change impacts may contain contested claims (i.e. incorrect or incomplete information) on some of the details of research findings. The effect of climate change on the Bay of Bengal cyclone (chapter 6) is an example of the importance of such details, as indicated above. Chapters on heat waves in Delhi and Toronto (chapter 7), climate change politics in USA (chapters 8) and climate change politics in Canada (chapter 9) are based almost entirely on discourse analysis of newspaper reports. These three chapters provide examples of how the media construct information on climate change by combining various angles of scientific information with pertinent climate-change policy issues. In these chapters, readers will find some examples of contested scientific information as well as some evidence of social amplification of risk.

Concluding planning implications

Notwithstanding afore-mentioned social amplification of risk in media discourse and political debates on climate change, some of the scientific facts on climate change are irrefutable. As numerical data and other information in chapters 2-5 indicate, global warming and climate change are real issues confronting us. Hopefully, this will motivate you/ us to take some actions for minimizing greenhouse gas emissions which have been responsible for the ongoing global warming and climate change. At least two types of actions are feasible on our part. **First**, at a political level, we can vote for politicians (at all levels, i.e. municipal, state/provincial/federal) who are known for supporting climate change legislations (for example, legislations for reducing greenhouse gas emissions). In addition to voting, we should be vigilant about post-election activities of these politicians so that we could pressure them to act on promised climate change legislations.

Second, we could do a lot at a personal level to reduce our carbon foot-prints which can be achieved by taking some of the measures to cut down our energy use and change our consumption behavior. In other words, we should all be aware of our carbon foot-prints and make smart choices in our daily life. The following are some of our suggestions:

Modify mode of transport:

- Instead of driving short distances, we should try walking or riding a bicycle.
- Use public transit to get to and from work.
- Car-pool whenever possible.
- Avoid commercial airline flights whenever possible.

Measures around homes

- Be sure to check the insulation in our home.
- The more efficiently we would use energy for heating and cooling, the lower would be our carbon footprint.
- Simple measures like sealing air leaks around windows and doors can have an enormous impact on our carbon emissions.

Eat local foods:

- We should try to eat local foods whenever possible.
- This will reduce the transportation component in the food supply.

Considering that we are so much accustomed to the comfort of modern life that is heavily dependent on fossil fuel energy, some of the measures suggested above may be challenging to follow. However, our view is that an informed consumer can have a meaningful impact on climate change!

References

Hannigan, J. A. 1995. *Environmental Sociology: A Social Constructionist Perspective*. London and New York: Routledge.

Hannigan, J. A. 2006. *Environmental Sociology*, 2nd edition (London and New York: Routledge.

Kasperson, R.E, Renn, O., Slovic, P., Brown, H.S, Emel, J., Goble, R., Kasperson, J.X. and Ratic, S. 1988. "Social Amplification of Risk: a Conceptual framework". *Risk Analysis* 8(2): 177-187.

Reckelhoff-Dangel, C. and Petersen, D. 2007. *Risk Communication in Action: The Risk Communication Workbook*. Cincinnati, OH: EPA Office of Research and Development, National Risk Management Research Laboratory.

INDEX

Absorption, of radiation, 20
Air pollution, defined, 24
Air temperature
 annual ranges, 99
 normal, 64
Albedo (reflection), 4, 19-20
Annex I countries, 29
Appliance and Equipment Standards Program, 165-166
Atmosphere
 composition, 14
 height, 12
Atmospheric window, 20

Buoys, temperature monitoring, 40

Canada, geographical settings, 207
Cap and trade (emission trading), 164
Carbon dioxide
 as greenhouse gas, 17, 25
 atmospheric concentration, 17
 sources, 25
Carbon foot-print reductions, 218-219
Carbon tax, 191-192
CFCs, 18

Climate Action Plan, Obama (2013), 165

Climate change
        activist organizations, 177-178
        and religion, 179-180
        as popular science (CCAPS), definition, 1
        causes, 215
        communication, 156-157, 216, journalists' role, 156-157
        debate, Al Gore, 215
        definition, 63
        denial organizations, 176-177, Republicans, 176-177
        forecast, Toronto, 120-121
        impacts, contested hypothesis, 63, 189 (letters)
        initiatives, US state governments, 174
        legacy, Obama's, 184-186
        opinion, George P. Bush, 155, Hillary Clinton, 156, Obama, 156
        policies, Trudeau's (Canada), 211
Climate sensitivity, 17
Climatic normals, 64
Cloud feedback, 4-5
CO2 level, data, 3
Coal, oil, and gas lobbies, 176
Coal reserves, USA, 162, Canada, 209
Coal-fired plants,
        on hold, 165
        opposition by Bush, 167-168
Coastal Mediterranean climate, 48
Cognitive claims by scientists, 8
Context issues, newspapers, 196
Continental climate, Saudi Arabia, 52, Canada, 47, 99
Cooling centre, Delhi, 110, Toronto, 119-120
Counter-radiation, 21
Coupled ocean-atmosphere phenomena, 84

Death, heat waves in India, 105-106
Dehydration, Delhi, 108
Discourse analysis, definition, 104, 149

El Niño events, 63, 84
Electromagnetic radiation, 11, spectrum, 12

Emission rates, per capita, 31
Emission reductions
       Annex I countries, 30-31
       Canada, 209
       public support, 190 (letters)
       state initiatives, 162-163
Encyclical, Pope Francis, 181; comments by Republicans, 182-183, 194-195 (letter)
Energy efficiency
       definition, 165
       local level initiatives, 167
       regulation delays, 166
ENSO, 86

Forest fire, California, 159-160
Frame analysis, 151

GISS, 43
Glacier retreat, Alaska, 159
Global average temperature increase, 3
Global commons, 23
Global warming
       anthropogenic hypothesis, 2
       contested assumption, 154
       debate, 153-154, 188 (letters)
       definition, 3, 63, 145
       evidence from India, 111
       IPCC projection, 214
       potential, 146
       trends, 44, 55
Greenhouse effect
       definition, 15
       newspaper review, 122
Greenhouse gases, anthropogenic, 11, 146
Greenhouse gas emission rates
       USA, 146
       Canada, 207

Heat capacity, 45
Heat waves (heat alerts, heat warnings)
        Canada, 68, 102, 112, India, 102-103, 107
        causes in India, 111
        climate change impacts, 97, 98
        definitions: Canadian, 102, Indian, 101-102, 110
        fatality rates, Delhi vs Toronto, 125
        health issues, Toronto, 116-117
        impacts, 97
        newspaper discourse, Delhi vs. Toronto, 125
Humid continental climate, Toronto, 100-101
Humid subtropical climate, Delhi, 99-100
Humidex, Canada, 113
Hyper arid climate, Saudi Arabia, 52

Infrared radiation (thermal radiation), 13, 21
Interpretive claims by scientists, 8
IOD, 86

Keystone XL Pipeline, 192-193 (letters), Canadian alternative, 208
Killer heat, Delhi, symbolic news, 112
Kyoto Protocol, 28, 161
        emission reductions, 161
        Canadian position, 208
        Republican opposition, 173

La Niña, 85
LOTI, 44

Marine climate, Vancouver, 48
McCain-Lieberman Climate Stewardship Act, 2003, 162
Media decontextualization, 8
Media discourse, 217-218

Media frames/news angles, 104
Methane
        global warming potential, 18
        natural and anthropogenic sources, 26

Milankovitch theory 5

Model uncertainty, 6
Montreal Protocol, 1987, 18
Most vulnerable country (MVC), Bangladesh, 49-50
Motor vehicle pollution control, California, 174-175

Natural gas-fired plants, 167
Next Gen (Tom Steyer), 178
Nitrous oxide (laughing gas), 19, 26
Nuclear power, 171

Ocean feedback, 5
Oil and coal lobbies, 215

Open system, atmosphere, 23
Orbital forcing, 5
Ozone, 27, dual role, 15

Paris climate summit, 2015 (COP 21), 186
Photosynthesis, 25
Polluters, global leaders, 29
Pollution units, data types, 28
Power saving devices, Delhi, 110
Prevailing winds, 83
Public discourse, 62

Radiation bands, 12
Radiation budget, definition, 19
Radiation, wavelength, 12
Radiative forcing
definition, 146
methane, 146
Radiative or blackbody warming, 3
Rainfall, desert climate, 52-53
Record high temperatures, Delhi, 106
Renewable energy, Georgetown, TX, 172, Massachusetts, Alaska, Texas, 171-172

Roast pit (symbolic news), Delhi, 111

Saffir-Simpson scale of hurricane/cyclone wind speeds, 81
Sampling
        judgment, 104-105
        limitations, 195
        random, Toronto heat waves, 114
Sandy, climate change link, 160
Scattering of radiation, 19
Science communication, 1
Sea ice, loss in Alaska, 158
Sea level rise
        Alaska, 170, New York, 169, Virginia, 168-169
        Atlantic and Gulf coasts, 158, Florida Keys, 155
Sea surface temperature (SST), 78
Social amplification of risk, 215-216
 Social constructionist perspective, 7
Solar forcing, 5-6
Solar power, Schwarzenegger's vision, 172
Solar radiation, definition, 12
Southern Oscillation Index (SOI), 86
Stay indoors, Delhi, 109
Stratosphere, 18
Subtropical high pressure, 52 (Saudi Arabia)
Sunspot activity, 6
Super-cyclones, 83
Sweats and heat strokes, Delhi, 108

Tailpipe emission reductions, California, 163
Tap water ban, Toledo, 159
Temperatures
        above average, 44
        annual averages, 65, USA, 147
        annual ranges, Bangladesh, 50, Canada, 47
        anomaly, 41, index values, 42
        changes in Saudi Arabia, 54
        long-term averages, 44
        maximum daily, 67, Delhi, 103, 107, Toronto, 68, 101

normal daily, 66
    seasonal averages, USA, 147-148
Temperature grid, 42
Temperature increases, average global, IPCC projection, 152-153
Temperature interpolation, 41
Thermal climate, 45, 71
Thermal radiation (thermal infrared radiation), 13
Troposphere, 18

UNFCCC, 28
Upwelling, 84
Urban heat island, India, 111
    Vapor feedback, 4
    Vapor, as greenhouse gas, 16

Volcanic eruptions, cooling effect, 5

Walker Circulation, 86, 90-91 (endnotes)
Warmest year, 40
Warming acceleration, 44, 56, Toronto and Vancouver, 49
Warming trend, Bangladesh, 50-51, Saudi Arabia, 53-54
Water vapor, as greenhouse gas, 24
Wind farm turbines, 171
Wind systems, global, 23

Printed in the United States
By Bookmasters